Politics, Geography and 'Political Geography'

A Critical Perspective

JOE PAINTER
University of Durham

A member of the Hodder Headline Group
LONDON • NEW YORK • SYDNEY • AUCKLAND

First published in Great Britain 1995 by
Arnold, a member of the Hodder Headline Group,
338 Euston Road, London NW1 3BH

Co-published in the USA by Halsted Press,
an imprint of John Wiley & Sons, Inc.,
605 Third Avenue,
New York, NY 10158-0012

British Library Cataloguing in Publication Data
A catalogue record for this book is available from the British Library

Library of Congress Cataloging-in-Publication Data
Painter, Joe, 1965–
    Politics, geography, and "political geography": a critical
perspective / Joe Painter.
        p.   cm.
    Includes bibliographical references and index.
    ISBN 0-340-63215-1 (hb.).—ISBN 0-340-56735-X (pb.).—ISBN
0-470-23544-6 (U.S.) : hb.).—ISBN 0- 470-23543-8 (U.S. : pbk.)
    1. World politics—1945– 2. Geography. 3. Geopolitics.
I. Title.
D843.P27 1995   95-21561
320.1′2—dc20   CIP

ISBN 0 340 63215 1 HB
ISBN 0 340 56735 X PB
ISBN 0 470 23544 6 HB (USA only)
ISBN 0 470 23543 8 PB (USA only)

1 2 3 4 5 95 96 97 98 99

Typeset in 10/12pt Sabon by
Phoenix Photosetting, Lordswood, Chatham, Kent
Printed by J. W. Arrowsmith, Bristol, UK

To my parents, Beatrice and Desmond Painter,
with thanks and much love

Contents

Preface

The initial idea for this book arose while I was at working at the University of Wales in Lampeter. I was teaching a course on political geography, which covered the usual topics of states, nations, territories and global and local power relations. Given the dramatic political changes which were going on in the world around me, this seemed highly topical, as well as allowing me to make good use of a long-standing academic and personal interest in politics. Casting round for ideas through which to interpret contemporary political change, I saw that many of the most interesting conceptual developments and theoretical debates in human geography seemed to be those associated with an upsurge of interest in social and cultural theory. The implications of these theories were widely regarded as highly political, and yet they seemed to be relatively little used by political geographers studying the kinds of substantive topics in which I was most interested. I am delighted to say that four years on, increasing numbers of geographers (and others) are drawing on the insights of social and cultural theory in explaining political-geographic change. The aim of this book is to show how some of these insights can help to make sense of the field of political geography.

There a number of people who have helped me along the way, and whom I would like to thank. Graham Smith first introduced me to political geography when I was a student and encouraged me to get this project off the ground in the first place. John Allen, Felix Driver and Miles Ogborn provided very constructive comments on a draft of Chapter 1. Although they did not always realize it at the time, I have had stimulating conversations on the subject-matter of the book with Parminder Bakshi, Paul Cloke, Mike Crang, Philip Crang, Mark Goodwin, Isobel MacPhail, Miles Ogborn, Chris Philo and Alan Southern. The Geography Departments at the University of Wales, Lampeter and the University of Durham have proved welcoming, stimulating and supportive environments in which to work. Laura McKelvie at Arnold has been patient and supportive. I owe most of all to Rachel Woodward, who has helped with ideas, discussions, constructive criticisms and much else besides. None of these people bears any responsibility for the shortcomings of the finished product. Finally, I have dedicated the book to my parents, Beatrice and Desmond Painter, with much love and thanks for their support over the years. They first showed me the importance of politics. I hope this book will show them why geography may be important too!

Introduction

> The next challenge for [political geography] is to incorporate new politicizations of geography through resurgent cultural geography and feminist geography.[1]

The breadth of material that might be covered under the heading of 'politics and geography' is vast and this book is not intended to be a comprehensive textbook or survey of the field. Nor is it a conventional textbook for the much narrower range of topics covered by the traditional sub-discipline of 'political geography', though I have focused on those topics in most of the chapters. Rather, it uses those topics to illustrate some contemporary developments in ways of understanding the relationship between politics and geography.

In recent years human geography has seen a considerable blurring of its constituent sub-disciplines. The traditional divisions between economic, social, political and cultural geography seem increasingly irrelevant as geographers focus more and more on the connections between them. It no longer makes sense (perhaps it never did) to think of separate economic, political and cultural 'spheres', each with distinctive geographical conditions and effects. Among others, Marxist geographers have charted the connections which make the economy 'political', feminist geographers have thought in new ways about the division between public and private which has undermined any essential distinction between the social and the economic and political, while environmental geographers have shown the importance of understanding the ways our use of, and impact on, the environment depends on particular cultural constructions of 'nature' and of our relationships to it.

One of the liveliest of these refashionings of the study of human geography has taken place in social and cultural geography. Among other things, this has involved a focus on a range of new concerns, including:

1 *The communication of meaning.* Drawing from work in cultural and media studies, geographers have become increasingly interested in the

ways in which social life is rendered meaningful to people. The process of ascribing meaning ('signification') is seen as an unequal one, so that different meanings operate to advance the interests of different people or social groups. Meaning is not seen as transparent and clear, but as socially produced and contested.

2 *The production and effects of discourses.* Related to 1, the concept of 'discourse' refers to a range of meanings, or meaningful statements, which come to be linked together in a broader framework. The framework, or discourse, provides a particular 'mode of thinking' which allows us to understand things in a certain way. For example, a discourse might identify some issues as more important than others, or some forms of behaviour as better than others. They are not (necessarily) more important or better in any absolute sense, but they are made to seem so by the discourse.

3 *Human subjectivity and identity.* Our 'subjectivity' is 'who we are', or rather 'who we feel ourselves to be' and 'who we are made to be' by society. Social and cultural geographers have been interested in the development of different subjectivities and identities in different places and among different social groups, and the ways in which the construction of identities happens through the operation of different discourses. Many writers have suggested that as individuals we all have multiple identities: we are different people in different contexts.

4 *Critique of geographical knowledge.* Geographical knowledge does not consist of transparent and value-free truths in the way that has often been assumed in the past. Like all knowledge it is the product of particular social and political contexts, and as such it advances certain interests, often at the expense of others. One aspect of the 'cultural turn' in geography has involved investigating the process through which geographical knowledge has been (and is) produced, and uncovering the (often unequal) power relations which it serves.

5 *The operation of human agency.* 'Human agency' refers to the capacities of human beings and their role in producing social outcomes. While human beings are not able just to do anything they please, human agency does 'make a difference', even if its effects are not always intended. Human agency is always situated in and conditioned by particular geographical contexts. Unequal access to resources and knowledge means that the capacity of some groups and individuals to make a difference is greater than that of others.

Although first highlighted in social and cultural geography, interest in these issues has recently spread much more widely through the discipline of human geography as a whole (not least because of the blurring of boundaries between its constituent parts). To date, though, that spread has been somewhat uneven. The sub-discipline of 'political geography' has only intersected with such concerns in a fairly limited way so far. This is rather ironic, since the connections between discourse, knowledge, meaning, agency and identity are highly political, at least with a small p. It might be argued that, given the blurring of sub-disciplinary boundaries, some unevenness in the diffusion of new perspectives is unimportant. If the boundaries break down, human geographers will feel free to adopt all kinds of perspectives in their investigations of all sorts of phenomena, without regard to the sub-disciplinary pedigree of either. While I am all in favour of such intellectual libertarianism, I also realize that the academic discipline of human geography is itself a social product with its own structures and institutions, which take time (and effort) to change and whose development often lags behind changes at the level of ideas.

One consequence of this is that although the issues of discourse, knowledge, meaning, agency and identity *are* very political, and have considerable implications for the way we understand what politics is, and how we study it, they have only rarely been applied to the study of the geography of Politics (with a capital P). This term refers to the institutions and processes of the state, government and formal political organizations which have in the past made up the subject-matter of the sub-discipline of 'political geography'.

Happily, moves in this direction are now getting started. The purpose of this book is both to present some of them to a wider audience and to contribute in a small way to the further blurring of the various sub-disciplines. My principal interest is in how we might best interpret and understand changes in the complex relationship between geography and politics. I am not concerned to present a full account of all the detailed shifts in the political geography (however defined) of the world around us. Rather, what I do want to do is to illustrate through selected examples what a political geography which is sensitive to social and cultural geography might look like.

Wherever possible I draw on the ideas and research of those geographers who have already made a start on this road. Where necessary, though, I also refer to the work of those in other disciplines if it illuminates the important issues. I hope the book will be useful and interesting not only to those taking courses in political and cultural geography, but also to all those concerned about the relationship between space, place and political processes.

The book is organized into six chapters. Chapter 1 discusses the characteristics of politics and presents the approach to studying politics which I will be using in the remainder of the book. The remaining five chapters deal in turn with some of the topics which have constituted the traditional subject-matter of 'political geography'. As I have suggested, while the topics are conventional, their treatment is intended to be less so. Chapters 2 and 3 examine the state. Chapter 2 considers the rise of the system of nation states and the process of state formation. Chapter 3 looks at the changing organization and role of the liberal democratic state in the late twentieth century and considers the claim that the state is in decline as a political force in the modern world. Chapter 4 covers imperialism and the continuing implications of colonialism for the relations between the West and its former colonies, and for the people who live there. Chapter 5 focuses on another aspect of the 'geopolitical world order': the startling transformations in global military and security relations following the Gorbachev reforms and the breakup of the former Soviet Union. The final chapter, Chapter 6 considers social movements and the participation of people in political change. Although often treated separately by political geographers, nationalist movements are included here as examples of social movements. Although I am quite deliberately not adopting an overarching explanatory framework, it will, I hope, become clear that there are very considerable interconnections between the political processes discussed in each chapter.

Note to Introduction

1 Peter Taylor, 'Political Geography', in Ron Johnston, Derek Gregory and David Smith, eds, *The Dictionary of Human Geography* (Oxford, Blackwell, 1994), p. 450.

|1|

Politics, geography and 'political geography'

What is this thing called politics?

Politics matters!

There is an ancient Chinese curse which, translated, runs: 'May you live in interesting times.' Cursed or not, from a political point of view life in the mid-1990s is certainly interesting. Great political changes have swept the globe in the past 10 years. In 1984, when I began studying geography at university, the world was in the depths of the new cold war. The hawkish Ronald Reagan had just been re-elected for a second term as President of the United States. In Moscow it seemed to be business as usual. Konstantin Chernenko emerged as the latest in a succession of Soviet leaders determined to remain true to the traditions of Soviet state socialism. The global military order organized around NATO and the Warsaw Pact was intact, with the United States in the process of deploying nuclear-armed cruise missiles in Western Europe in the name of 'collective security'. As far as domestic politics was concerned, the radical right-wing doctrines of monetarism and free-market economics that had been enthusiastically adopted by the Reagan and Thatcher governments in the USA and the UK had yet to generate much support in other industrialized capitalist countries. In China, just 10 years on from the end of the cultural revolution, economic reform was well under way, but there was as yet no hint of the lengths to which the Communist Party would go to maintain political control. While Eastern Europe, including the then Yugoslavia, remained politically stable, in other parts of the world, civil unrest or outright civil war were the order of the day. In South Africa, with no sign that Nelson Mandela would be released from prison, the struggle against apartheid was intensifying. In Central America the US-backed military campaign against the reformist Sandinista government in Nicaragua was in full swing.

Ten years on it seems as if the world has been turned upside down. As I write this on the last day of 1994, Nelson Mandela is President of South Africa following the dismantling of apartheid and the ending of white minority rule. The USSR no longer exists; its reforming last leader, Mikhail Gorbachev, is no longer even President of Russia. In Eastern Europe, which for long seemed to many to be not just stable, but positively stagnating, ethnic conflicts (and in places outright civil war) have caused the dramatic fragmentation of the political map. The Warsaw Pact has been dissolved and it is not impossible that Russia will eventually join NATO.

For a few brief months after the fall of the Berlin Wall there was serious talk about the possibility of a 'new world order', secured through the political resources of the United Nations and backed by the United States armed forces as a kind of self-appointed global policeman. The multiplication of a range of military conflicts and humanitarian 'disasters', which the combined forces of the UN and the US seem powerless to solve, has led some to speak unkindly, but probably rather more realistically, of a 'new world disorder'. The privatizing radicalism of the Thatcher governments caught on throughout Western and, especially, Eastern Europe, despite the departure from office of Mrs Thatcher herself. Meanwhile, in Tiananmen Square it became dramatically clear that free-market reforms in China were emphatically not to be extended to the liberalization of political life. In Central America the Nicaraguan people, tired of the effects of continual US attempts to destabilize the Sandinista government, voted it out of office.

Political change has certainly been profound and dramatic. Yet it has also been paradoxical. Politics has never been so important, yet, at the same time it has never been so unpopular. Scarcely a week goes by without a newspaper article or a television feature on the distrust and contempt with which the public in many countries regard their elected representatives. Part of this stems, no doubt, from the personalities involved. However, there seems to be something deeper going on. There seems to be an increasing perception, in the West at least, that governments simply no longer have the power to influence events in the way that they once did (or at least claimed to). Governments frequently claim that global economic forces are shaping their national economies, and there is relatively little that they can do to intervene. While this is partly used to explain the failure of policy, it is certainly the case that economic processes do flow across international boundaries as never before, limiting the capacity of any one government to affect their direction. This has led to new kinds of relationships between governments as they try to regain control. The growth of the European Union and the setting up of a North American Free Trade Area, for example, are, in part, attempts to exert political influence over economic affairs at a wider geographical scale than that of the nation state.

At the same time, however, such 'supra-state' institutions raise political questions too. In a free-trade area, not all regions or countries benefit equally, and the European Union has had to establish special funds to support those regions where exposure to international market forces would cause major social upheaval. In addition, larger political institutions are perceived by some as a threat to national or regional cultural distinctiveness. Some of these questions will be examined in more detail in later chapters.

Traditional political divisions and organizations, which have provided the stable framework for political debate, participation and policy making in so many countries for half a century or more, continue to endure. But to many, they seem to be increasingly out of touch and inappropriate as their traditional constituencies are altered, sometimes dramatically, by successive waves of social and economic change. New political organizations with leaders like businessmen Silvio Berlusconi in Italy and Ross Perot in the US arise with dramatic popular support, often to fade away as quickly as they came. In many places the difficulties of coping with economic problems and a shifting political landscape have led people to make scapegoats of some of the most powerless in society, rather than to seek reform and development through conventional political channels.

On the other hand, social idealism is far from dead, with a wide range of groups and individuals seeking to mobilize around specific issues such as environmental protection, civil rights for disadvantaged social groups, and the provision of adequate and appropriate health care, sanitation, education and means of making a living to the three-quarters of the world's population who currently live without them. Here too, though, frustration with conventional politics is common, as campaigners experience firsthand the difficulty of producing policy changes which really make a difference in these areas.

Thus the paradox is that in a time of extraordinary political transformation there is apparently such widespread scepticism about, and even downright mistrust of, the formal political system. Perhaps, though, this is not really paradoxical at all. Perhaps it is *because* of the social and economic instability of the contemporary world that familiar political traditions, systems and ways of thinking have come to seem increasingly irrelevant. No leader, party or political movement seems able to find a language (still less a set of policies) which capture the spirit of the age. As the British political thinker and writer Geoff Mulgan says,

> beneath the inertial momentum of elections and offices, the political traditions that became organizing principles for so many societies, dividing them into great tribal camps identified with class, with progress or reaction, with nation or liberty, have lost their potency. They cannot inspire or convince. They do not reflect the issues which passionately divide societies. They are no longer able to act as social glues, means of recognition across distances of geography and culture. What remains is a gap, psychic as much as instrumental. Without great movements, it is much harder to understand your place in society, much harder to picture where it is going. And without coherent

political ideas, to organize the fragments of many issues, fears and aspirations, it becomes far harder to act strategically and to think beyond the boundaries of individual lives and relationships. It is not that the great questions have been answered: just that the available solutions have lost their lustre.[1]

Yet politics is not just going to stop. The range of issues and problems facing us seems destined to grow, rather than shrink. Environmental change, health and disease, military conflicts, economic problems, ethnic identity, cultural transformations, global poverty: the list seems endless. However we deal with (or neglect) these concerns we will be engaging in politics. Directly or indirectly politics permeates everything we do and influences all our lives. Politics matters!

Politics formal and informal

Most of us, I suspect, operate most of the time with what we might call a 'common-sense' or 'everyday' understanding of what politics is. In the common-sense view, politics is about governments, political parties, elections and public policy, or about war, peace and 'foreign affairs'. All of these are immensely important, and I will be discussing them (indeed concentrating on them) in this book. However, these common-sense assumptions about politics are rather limited. They refer to what I shall call 'formal politics'. By 'formal politics' I mean the operation of the constitutional system of government and its publicly defined institutions and procedures. The implication is that politics is a separate sphere of life involving certain types of people (politicians and civil servants) or organizations (state institutions). The rest of us interact with this separate sphere in limited and usually legally defined ways. The political system may accord us formal political rights (such as the right to vote, or to own property) or formal political duties (such as the duty to serve on a jury, or to pay tax). Alternatively it may from time to time affect the society in which we live, through changes in public policy, for example in the spheres of education or environmental protection. Most of the time, though, many people do not think much about formal politics. Because it seems to be a separate sphere, we can say things like 'I'm not interested in politics' or 'he's not a very political animal'. Formal politics is seen as something that can sometimes *affect* everyday life, but is not really *part* of everyday life.

One thing I hope this book will show is that the formal political system has much more impact on our lives than is often realized. Of course, the extent to which society is openly controlled or influenced by the government varies considerably. In some countries (such as those still governed by absolute monarchies, for example, or various forms of central planning) the presence of the government in daily life may be clear and explicit. However, even in the so-called 'liberal democracies', the role of the state and the formal political system is wider and deeper than the notion of a separate and limited political realm would suggest. The difference is that in more 'liberal' societies it is easier to believe in the separateness of formal politics, because its presence, though significant, is either hidden, or taken for granted and unquestioned.

By contrast, 'informal politics' might be summed up by the phrase 'politics is everywhere'. A good example is the idea of 'office politics'. Office politics obviously does not have much to do with the political system of governments and elections, but everyone understands why we refer to it as 'politics'. It is about forming alliances, exercising power, getting other people to do things, developing influence and protecting and advancing particular goals and interests. Understood like this, politics really does seem to be everywhere. There is an informal politics of the household (parents attempt to influence children, women do more housework than men); of industry (some groups of workers do better out of industrial change than others, the aims of management and workers often conflict); of education (some subjects and points of view are taught while others are not, some children benefit more from education than others); even of television (some people have more chances to have their say on TV than others, certain groups are shown in a more favourable light than others). In fact, if we are talking about *infor*mal politics, there is no aspect of life which is *not* political: politics really is everywhere.

It is often said that 'politics is about power'. This statement raises rather more questions than it answers, and I will be discussing some of them in more detail later in this chapter. However, I want to consider it briefly here in relation to the ideas of formal and informal politics. The ways in which power has been understood by social scientists have changed over time. According to the French philosopher Michel Foucault, such changes are related to changes in the ways power is exercised. Foucault argues that in traditional societies power was exercised visibly, in, for example, public spectacles. It often took the form of dramatic acts or displays. In modern societies, by contrast, the exercise of power is much more hidden. To take one of Foucault's own examples, in the punishment of criminals the power of the state in traditional (medieval) societies was displayed through theatrical executions in a public square. These practices gave way during the eighteenth and nineteenth centuries to what Foucault calls the 'disciplinary society'. In the modern disciplinary society, he argues, social control is produced by a complex network of rules, regulations, administrative monitoring, and the management and direction of people's daily lives. This is most strongly developed in institutions and organizations such as prisons, schools and factories. To some extent it applies to all our private lives as well. In the disciplinary society, we all, to a greater or lesser extent, internalize codes of behaviour and rules of conduct, so that we are unconsciously disciplining ourselves. Instead of being dramatic, public and visible (as in traditional societies), power in modern societies is invisible. It operates behind the scenes, as it were, and in every part of the social order. Instead of something which exists in the centre of society (with the king, say, or the government) and which is consciously used against the powerless, power now flows through all the complex connections of everyday life. Foucault's concept of power in modern societies is sometimes referred to as a 'capillary' notion of power, to imply that power filters down through all our most mundane and ordinary relationships and out into the most routine aspects of human activity.

Foucault's ideas about power are controversial, and I have, in any case, simplified them sharply. However, it seems to me that there is an interesting parallel here with the notions of formal and informal politics. From a Foucauldian perspective, the claim that 'politics is about power' takes on a particular meaning. If power in modern societies saturates the social fabric in the ways Foucault implies, then studying politics should involve at least as much emphasis on informal politics as on formal politics. Moreover, the capillary notion of power implies that power, and hence politics, is part of all social life and all forms of social interaction, however normal, mundane and routine they seem. Thus the way we feel about ourselves and others, how we write and talk, how we work and shop, how we study and play, how we drive and go on holiday – all of these are political, as are our religious, recreational, sexual, artistic and academic activities. This is somewhat unnerving, to say the least, and many people may be unhappy to think that their private lives have anything to do with politics. However, if by 'private' we mean not affecting, or affected by, other people or organizations, it is remarkable how little of modern life can be counted under that heading. Almost all the areas of daily life I have mentioned are likely to involve other people to some extent, even if indirectly. When you shop for food, who grows it, under what conditions and how much are they paid? When you go on holiday, what effects do you have on the places you visit and the people who live there? When you write, what kinds of expressions do you choose to refer to other people, and what kind of representation do you build up of them? We may not feel (or may choose not to feel) responsible for the people with whom we have these indirect relationships,[2] but like it or not, we *are* involved with them.

A DILEMMA

The idea that power relations and politics exist in all social relationships, and not just in the institutions of the state and the government, is an important insight. However, it also raises a couple of questions. One is practical. Given that everything is political, and that no one book can cover everything, what should be included in a book on politics and geography? The other question is conceptual. The 'politics is everywhere' approach is useful in pointing out the limits and shortcomings of the traditional view of 'politics as government'. On its own, however, it makes it difficult to decide which elements of politics (which types of social relationships) are more significant or important than others. There seems to be a risk that 'politics' just dissolves into 'everyday life'.

My answer to the first question is partly a pragmatic one. The topics I have chosen to cover in this book are the ones which make up the field of study of 'political geography'. Despite its name, the sub-discipline of political geography has not, in the past, dealt with the full range of 'politics' which I have been talking about. It has usually concentrated on formal politics, and even then on particular aspects of formal politics: a mixture of those that were commonly studied by the founders of the subject, and those that were regarded as being somehow especially 'geographical'. (Today it is becoming clearer that virtually all political processes are geographical, although not always in obvious or conventional ways.) However, as well as being a pragmatic choice there is another reason for selecting these topics. The larger field of informal politics is also being widely studied by geographers. In recent years they have introduced a whole set of new ideas into geography drawn mostly from social and cultural theory, which I believe is particularly useful in thinking about what politics is and how it works. On the whole, however, geographers looking at these broader notions of politics (the politics of everyday life, and so on) choose not to call themselves 'political geographers', and choose not to apply their ideas to the formal politics usually studied under the heading 'political geography'. In this book, therefore, I aim to show how some of the theoretical perspectives which have helped to illuminate informal politics may assist us in understanding formal politics.

The temptation raised by the second question is to think that if politics is everywhere, then it is not really possible to analyse or interpret it as a distinct aspect of human life at all. My answer to this problem requires me to spell out in a bit more detail how politics works, and the framework through which I want to think about politics in the rest of the book.

The nature of politics today: an interpretative framework

Politics as social practice

Above all, politics is about people and their relationships to others. Most people, most of the time, like to think of themselves as individuals. We want to be free, autonomous and capable human beings, not subject to the whims and power of others. In the West this dominant common-sense view of life is based on the doctrines of 'liberal individualism'. We are brought up to believe in this common sense through what we hear from parents, schools, the media and politicians. This dominant idea is a set of beliefs, rather than an accurate picture of society, but it has been made to seem the *natural* way of things, despite the fact that it arose in specific *social* circumstances.[3] Some of the time our experiences bear it out, most notably when we are able to achieve things we want to do. More often we find we cannot achieve what we want to do, at least not entirely. On these occasions we may blame our-selves. Perhaps we could have done it, if only we had worked harder, or been cleverer, or in some way better. Some of the time it may indeed be our 'fault'. Often though, we are unable to get our way because of others. We think of ourselves as free, but our freedom is partial, limited and dependent on other people and organizations. This is my starting point for outlining an interpretative framework through which we can try to understand and explain political change.

The interpretative framework I want to use sees politics as made up of social practices. Social practices are usually both *material* and *discursive*. The material aspects of social practices are those which involve the organi-zation and use of things. The discursive aspects are those which involve ideas, language, symbols and meanings. Thus eating a meal, for example, involves material practices (the preparation of foodstuffs) and symbolic or discursive ones (an understanding of the role and meaning of meals and mealtimes in society). Writing a book involves material elements (paper, pens, word processing, the printing process) and discursive elements (the ideas in the book, the significance of literature as a cultural form and so on).

While material practices and discursive practices can be distinguished for the purposes of analysis, they cannot exist independently of one another. For matter to be used by human beings, they must have a (discursive) under-standing of its role and importance. Equally, discourse is produced materi-ally (whether it involves thought, speech, writing, graphics or takes some other form, the form always exists as matter), and (often) has material processes or practices as its subject-matter. The material cannot be sepa-rated from the discursive, but they are not the same thing. It is common for different writers on politics and geography to emphasize material processes over discursive ones, or vice versa. I want to argue that neither can be under-stood in the absence of the other.

During the 1970s and early 1980s, critical human geography was domi-nated by an emphasis on political economy, usually drawing on interpreta-tions of the work of the nineteenth-century social thinker, Karl Marx. Many of these formulations gave implicit or explicit priority to material processes in the 'mode of production'. From the mid-1980s onwards, these approaches were supplemented (and sometimes challenged) by a growing group of critical human geographers who, in various ways, stressed the role of discursive and cultural practices in the development of social life.[4] In my view both these sets of insights are important, but both are partial. Human life is both material and discursive, and the more we investigate the complex relations between the two, the more difficult it becomes to accord a general primacy to one or the other. This is perhaps particularly true of politics, which, as I outline it below, involves both material interests and discursive argument, both 'modes of production' and 'discursive formations'.[5]

MODE OF PRODUCTION

The concept of 'mode of production' refers to the ways in which individuals and social groups are provided with the means of fulfilling their needs and wants, from biological necessities such as food, to the most frivolous luxuries. In complex, modern societies, the mode of production is correspondingly complex. Drawing on ideas from political economy, we may identify a number of key elements. First, the process of production requires the means of production (offices, computers, machines, tools, factories and so on), raw materials and human labour power. Second, the process of production is organized in different ways in different times and places. In craft production, for example, the labourers own the means of production themselves. Under capitalism, the ownership and control of the means of production is separated from the direct producers. Third, there is a division of labour, through which different parts of the production process are allocated to different social groups. Fourth, there is a system of distribution or circulation through which products can be allocated to consumers. These various elements and the relations between them take different forms in different modes of production. The social outcomes (who gets what, where and how) are usually systematically unequal, although the character and causes of the inequality are different in different social systems.

DISCURSIVE FORMATION

Like the 'capillary' notion of power, the concept of 'discursive formation' comes from the work of Michel Foucault. According to Foucault, the meaning of language is not transparent and immediately obvious. Words, statements, symbols, metaphors and so on, mean different things in different contexts. The meaning of a particular statement depends partly on who is saying it and how it is being said, but also on how it fits into an existing wider pattern of statements, symbols and understandings. It is this wider pattern which Foucault calls a 'discursive formation', which is often shortened in the work of subsequent writers to simply 'discourse'.

This will be clearer if we consider a simple example. All human beings who live to maturity pass through the ages of 13 to 19. However, it was only in the 1950s that the term and concept of the 'teenager' became widespread. Before then, in some societies, one moved more or less directly from childhood to adulthood. In others, such as in Victorian England, the term 'juvenile' was often used, but crucially it did not mean the same nor have the same connotations as 'teenager'. In most societies, specific ages or ceremonies were important in marking the transition from childhood, often at a customary age such as 21 (the age of majority in nineteenth-century England) or a religious rite of passage such as the Jewish bar mitzvah. In America during the 1950s, however, the stage between childhood and adulthood emerged as separate and was labelled 'teenage'. Human beings were still biologically the same, and yet Western society was transformed by the emergence of the 'teenager'. In Foucault's terms this was the result of a 'discourse of the teenager'. The 'statements' which made up the discourse were indeed dispersed throughout society. They appeared in many different media: in political speeches, in films, in popular music, in advertising, in newspaper columns, in parental discussions and so on. However, they all had enough in common, in their object of analysis, in their mode of language, in the terms used and their tone, to be considered part of a unified 'discourse'. The 'teenager' therefore, was a 'discursive construction' which was 'made real' by the discourse. While it referred to the same span of years as the Victorian concept of 'juvenile' the effects of the two discourses were very different.[6]

THE DISCURSIVE AND THE MATERIAL

Although some subsequent writers have tended to emphasize the role of discourse above (and sometimes to the exclusion of) all other factors, Foucault himself stressed the importance of 'extradiscursive' phenomena.[7] He insisted, however, that it is not possible to say which is more fundamental in the process of social change. For Foucault 'things' are made meaningful (and thus constituted *as* things) through discourse. Thus another way of looking at the concept of discourse is to see it as a framework within which things are made meaningful. The concept of the teenager would have been simply meaningless in the nineteenth century. It only makes any sense in the context of a specific discourse.

Throughout this book I will be stressing the importance for politics of the relationship between discursive and material practices. This is to move a little way from Foucault's ideas, as I want to discuss not only the role that discourse plays in making things real, but also the role that material practices play in enabling or constraining discourse. For instance, to pursue the teenager example, extradiscursive processes and practices seem to me to have played a considerable part in enabling the discursive construction of the teenager to make sense. These include the growth of the American economy, which provided the wealth and resources for clothes, records and cars; the availability of leisure time and of extended education; changing demographic and family patterns; and the construction of a material geography in American cities of coffee bars, movie theatres, shops, sports facilities and high schools. Although it may not be obvious at first, the same link between discursive changes and material conditions are significant in most areas of what we call politics. In the other chapters of this book I hope to show how political discourses are simultaneously the producers and the products of material practices in a process of mutual constitution.[8]

A framework for interpretation[9]

Having considered the notions of practice and the relationship between the discursive and the material, I am now in a position to outline the elements of the interpretative framework. In this framework I define politics as a process. This process has a number of features. First, it is constituted through geographically and historically situated social and institutional practices. Second, those practices are both material and discursive in character. Third, they are, at least in part, purposeful and strategic. Fourth, political practices depend on the availability of unequally distributed resources. Let me explain this definition in a little more detail, by outlining six key elements of the interpretative framework which will inform the rest of the book.

(1) PEOPLE AND THEIR COMPETING NEEDS

People and the relationships between them make politics; political processes are produced by human activities and human agency. As human beings, we all, individually and in social groups, have needs, desires, wants and interest which, with the possible exception of basic biological necessities, are constructed (made meaningful) through discourse. Politics arises from impossibility of reconciling the needs, desires, wants and interests of all individuals and groups instantly and automatically.

(2) THE ROLE OF STRATEGIC ACTION

We develop and pursue strategies (purposeful practices) in support of (our understanding of) our interests. Strategies need not be grand or comprehensive; they may be mundane or small scale. Our strategies are never wholly rational, since our knowledge of the circumstances in which we act is always partial and imperfect, and many of the factors which influence outcomes are beyond our control. This means in turn that while our strategies have effects, their effects are often unintended. Strategic action potentially brings actors into conflict or alliance with others pursuing similar or opposed strategies, and can consequently generate both struggle and co-operation.

(3) RESOURCES AND POWER

The ability of different groups and individuals to pursue strategic action varies, as does its effectiveness, depending on the differential availability of resources within society. Resources may be of many kinds. These include: our bodies; other material resources of all sorts; discursive resources (such as knowledge, information, language, symbols, and ways of understanding); the compliance of other people; means of violence; and organizational resources (the ability to co-ordinate, deploy and monitor other resources). Unequal access to such resources accounts for differences in political power. Where conflicting strategies are being pursued, the exercise of political power generates resistance (counter-power).

(4) INSTITUTIONS

Strategic action often leads to the development of institutions of various sorts. Once established, though, institutions escape from the intentions of the initial strategy and develop independently. Institutions are then political actors themselves, pursuing strategies that may be unrelated to those which established them. Institutions also have their own internal politics, which consists of individuals and groups also pursuing strategic action. The strategies of institutions are the (often unintended) products of internal politics. As such they may be (and often are) contradictory. Institutions exist on a different temporal (and often spatial) scale from individual action. The fact that they endure over time and are stretched over space is one source of their political power, and helps to explain why how they can become harnessed to very different strategies from those intended by their creators.[10]

(5) AUTHORITY AND SOVEREIGNTY

Individuals, groups and institutions typically advance claims to authority, through which they aim to secure the compliance of other individuals, groups and institutions with their own strategic action. However, there are no absolute grounds on which authority can be justified. All claims to authority are assertions, rather than statements of fact. Claimants to authority usually pursue (often again through strategic action) attempts to legitimate their assertion: that is, to secure consent to their claim from both other claimants and those whose compliance to authority is sought. The process of legitimation is a discursive one involving attempts to construct frameworks of meaning through which authority is made to seem legitimate. Legitimation is rarely completed or absolute, but is a continual struggle against those who contest it. In the absence of (or additional to) consent, compliance with claims to authority may be pursued through coercion, where the necessary resources (means of violence) are available. A claim to sovereignty on the part of an institution is a special type of claim to authority: a claim to being the highest authority for some defined group or area. Like all claims to authority it is rarely established and uncontested.

(6) POLITICAL IDENTITIES

Our pursuit of different strategies and our positions in relation to the strategies and claims to authority of others, constitute us in a variety of ways as political subjects with particular political identities. These are thus partly the products of our conscious intentions, but partly the outcome of the discursive and material practices of others. To say that we are political subjects means that we each, as human beings, have relationships to politics. Part of 'who we are' is produced through our political positions. For instance, we all relate to the state in different ways, perhaps as voters, as users of public services, as asylum seekers, as pupils in state schools or universities, or as the focus of various forms of legal regulation. In different times and places we take on different political identities, sometimes deliberately, as part of a strategy, and sometimes unwillingly or even unconsciously. The notion of political identity will be considered in more detail in Chapter 6.

A way of thinking

This may seem over-elaborate, but I want to avoid giving the impression that there can be a neat definition of politics along the lines of 'politics is about power'. The difficulty with a short definition like that is that it begs more questions than it answers. My rather more detailed formulations show that while in one sense politics really is everywhere and in every social situation, it is possible to identify what it is about a situation which is 'political'. Do not worry if this seems a little abstract at the moment. In the chapters that follow I will show how this kind of perspective can help us to understand political change in different contexts, which should help to flesh out the framework in more detail. It is important to note, however, that this perspective is not a rigid theory which can be applied like a template to all political situations. Rather, I want to use it as a *way of thinking* about politics. There will be times when I use other, more detailed theories to talk about particular aspects of political change. There are, for example, substantive theories of international relations, imperialism, state formation and social movements. In the chapters dealing with those topics, I will want to discuss some of those more specific theories and their strengths and weaknesses, using the above framework as a kind of guide for assessment.

Formal and informal politics and organizational culture

As I explained above, in the rest of the book I will be concentrating mostly on topics which have been the traditional focus of the sub-discipline of political geography. This means that I will be particularly concerned with what I have called 'formal' politics. However, from the perspective of my interpretative framework, formal and informal politics are not distinct realms: they overlap. All formal political institutions are subject to internal (informal) political processes. At the same time, the kind of strategic action typically involved in informal politics is often both regulated by formal politics and intended to influence it (through, for example, political campaigns and social movements). Formal political institutions are not built only around strategic action, however. First, they are the products of bureaucratic forms of organization which structure their day-to-day activities and provide for continuity over time. Second, they are also cultural phenomena. Meanings, symbols and discourses are not only features of 'everyday life' but are also central to formal politics. The concept of 'organizational culture' is an increasingly familiar one in both private and public sector organizations. Much of the writing on organizational culture treats culture as (potentially) homogeneous throughout institutions. The framework I have outlined suggested that this is unlikely to be achieved: there are likely to be competing and conflicting cultural values and symbolic and discursive practices within organizations, but that probably increases rather than decreases the significance of their cultural aspects.

To give a simple example, in public organizations, managers are nowadays giving much more attention to their 'corporate identity' than has been the case before. Critics often argue that a concern with logos, typefaces and glossy publicity is an irrelevant waste of resources in a time of financial stringency. Often, however, the underlying objection is that the public services are becoming more like private businesses. In that process, the development of corporate identities is not incidental, but a part of what makes the change happen. Although supporters of public service provision often criticize a concern with logos or corporate image as a frivolous distraction from the 'real' function of the organization, the vehemence of the criticism reveals the significance of the symbolic aspects of institutions. The conflict is not about whether institutions could ever be image-free or symbolically neutral (they clearly cannot); it is about the nature and meanings of the institution's image and symbolism.

The paradox of modern politics

At the beginning of this chapter I suggested that politics had become a rather frustrating affair, and something of a paradox. On the one hand, political change seems to be accelerating and affecting our lives more and more. On the other hand, our old ways of thinking and doing politics seem to make less and less sense. The power of governments seems to be waning, and many people regard the traditional forms of political representation as inappropriate or even part of the problem, rather than part of the solution.

Much of this frustration with conventional politics stems, I think, from the gap between our assumptions about the power and status of individuals and our experience of the complex social interactions I have outlined above. If our goals, aspirations and strategies (to say nothing of those of our governments) are so dependent on circumstances beyond our control, or even awareness, it is unsurprising that the outcomes of politics should so often fall short of our intentions. Yet what I have called 'strategic action' *is* worthwhile. Although our knowledge and other resources may be limited, they are not completely negligible, particularly if they are harnessed to those of others with whom we may make common cause. It may need new forms of political activity, but given the problems facing the modern world, politics not only matters; it matters more than ever before. Within these changes geography is of vital importance, and it is to these I now wish to turn.

Politics and geography

So much for politics: what about geography? Of course, the term 'geography' can refer to quite a wide range of ideas. Traditionally geography has been defined broadly as the study of the earth's surface. As far as human activity is concerned this is often thought to involve four (overlapping) aspects:

1 *Space*. Geographers study the spatial distribution of human activities and institutions of all kinds and their causes and effects. They are also interested in the influence of spatial organization on social, political, economic and cultural processes.
2 *Place*. Geography involves the study of place: the character of places, the relationship between people and their places, and the diverse role of places in human activities.
3 *Landscape*. Geography focuses on the development of landscapes and the meaning and significance of landscapes for people.
4 *Environment*. Geographers are interested in the relationship between people and their environments, including their understandings of environments and their use of environmental resources of all kinds.

All of these traditional concerns remain central to human geography in the 1990s. All of them, however, have been subject to considerable rethinking and reformulation over the past 25 years. To take one example, the relationship between society and space has been the focus of much debate within human geography. In the past it was often assumed that space and society were separate things which may have *influenced* each other in various ways, but which could in principle be examined and analysed independently. More recently many geographers have insisted that spatial relations are *inseparable* from society. All social relations are constituted spatially, and there can be no possibility of a non-spatial social science.[11]

In political geography, it is the relationship between space and political processes which is the focus of attention. Political geographers in the past have argued both that politics sometimes has geographical effects (such as the uneven outcome of policy decisions) and that geography can have an impact on politics (such as the role of geographically defined constituencies in electoral systems). Both these formulations capture something of the relationship between politics and geography. However, both of them are, to my mind, based on the same flawed assumption, that 'politics' and 'geography' are autonomous spheres of life which interact. If the idea that social relations are always inherently spatial is applied within political geography too, then the assumption that politics and geography are separate interacting phenomena becomes untenable. This implies that there can be no politics which is *not* geographical.

To understand what this entails consider some of the components of politics outlined above. Human agency and strategic action is always situated in particular geographical contexts, which condition strategies and make some options available and others impossible. Agents draw on resources in developing strategies, but the availability of those resources is determined in part by their spatial organization. In addition knowledge, information and symbolic understandings are the product of geographical contexts and often have places and geography as their subject-matter. Moreover, space and spatial organization is itself a resource. Studying the control of key sites and territories has a long history in political geography, but the principle may be extended much further. For example, the spatial organization of institutions such as schools, factories and prisons is a central element in their control and monitoring.

Finally, social and cultural geographers increasingly stress the ways in which the production of (political) subjectivities and identities is bound into space and place. All of this suggests that, given the understanding of politics I have outlined, geography is not an optional extra, or a particular perspective. On the contrary, politics is always constituted geographically, and needs to be studied in that light.

'Political geography'

Disciplines and sub-disciplines

The academic subject of human geography has traditionally been divided into a number of so-called 'sub-disciplines': urban geography, social geography, historical geography and so on. Given the breadth of the subject of human geography, this is unsurprising and in many ways eminently practical. Each sub-discipline notionally represents a more limited and specialized field of study supposedly corresponding to a coherent part of 'geographical reality'. However, universities, where academic human geography is mostly practised and developed, are not the rational, ordered places they sometimes like to claim to outsiders. Although some writers have tried to argue for the development of a rational sub-disciplinary structure,[17] the activities of human geographers seem unlikely to fall neatly into ordered categories. This is because the subject of human geography has evolved historically, and has been created in particular social and political conditions. For this reason, some sub-disciplines are stronger than others, with more research, more academics working in the field, more conferences, books and papers and so on. More significantly, academic research and writing is continually developing and changing. Understandings of the world and the subject form and

(and a key producer) of the discourse of a sub-discipline is its textbooks. In 1993 two political geography textbooks were published, both of which are widely used in English and North American universities.[13] Neither of them makes reference to women, feminism or gender in their indexes.[14] It is important to note that geographers are working on these topics, indeed they represent some of the livelier and more innovative areas for contemporary geographical research. For the most part, however, the resulting papers and reports have not counted as political geography according to the (unwritten) rules established by its discourse. This highly political work tends to appear in journals and books devoted to cultural or social geography, or those dedicated to studies of gender. Of course, in one sense this may not be important. If the work is done, and it is published, then perhaps it does not matter whether we call it political geography or cultural geography.

Fair enough. There are, though, two reasons why such turf disputes may be rather more important. First, the example illustrates the extent to which the conventional academic divisions do not reflect a rational ordering of reality (or even their own definitions of their fields) but a particular path of intellectual, social and institutional development. Second, as I mentioned above, discourses are the products of, and in turn sustain and promote, particular social and political interests. Discourses involve mobilizing meanings in association with relations of power. If the discourse of political geography constitutes the sub-discipline around (among other things) the geopolitical world order, while simultaneously marginalizing issues of gender and the politics of the women's movement, then the inference that students, policy-makers and other academics are likely to draw is that geopolitical transformations are not connected to (or even that they are more politically significant than) gender politics. Arguably, such a discourse implies that geopolitical change is not even of concern to women: that it is men's work.

In part discourses are defined by what they exclude: by the ideas which fall outside their frameworks of what is meaningful. In part, though, they are defined by the elements they include. Like other discourses, political geography is marked by its origins. Early political geography was concerned mainly with the relationship between physical territory, state power and global military and political rivalries. In large part, contemporary mainstream political geography continues to take these concerns as its starting-point, and they are, of course, important issues. Nevertheless, the result is a rather top-down view of the subject.

Until recently, *people* appeared rather infrequently in the textbooks of political geography. Political geographers now take social and political struggles and social movements much more seriously than in the past, but these concerns are often coloured by the historic discourse. For example, the social movements which political geographers study more frequently than any other are those associated with regionalism and nationalism. These movements are distinctive in being based explicitly in an appeal to territoriality. By contrast, although other social movements, such as those associated with labour, women, gay and lesbian rights, minority ethnic groups and the environment, are all highly political and all exhibit marked geographical characteristics, they have been until recently largely neglected by political geography because they do not fit so well with its traditional themes.

The dominance of a particular way of thinking in mainstream political geography is shown in the following quotation from one of the textbooks mentioned earlier. In his discussion of imperialism, Martin Glassner cites 'cultural imperialism' as another form of imperialism along with military imperialism and economic imperialism. Cultural imperialism, he avers, 'is still more subtle, sometimes less effective, and *certainly less geographic* than the other forms'.[15] One of the liveliest sets of debates within contemporary geography in recent years has been concerned precisely with the deep and complex relationship between culture, imperialism and geography. This suggests that what Glassner has in mind by 'geographic' is the control of physical territory – an interpretation which stems directly from what I have been calling the discourse of political geography.

No discursive formations are unchanging, however, and in the last few years the sub-discipline of political geography has become open to a much wider range of topics and perspectives. In the early 1980s, a marked resurgence in the subject occurred. While many of the topics studied remained much the same, political geographers' investigations of them began to draw heavily on ideas and perspectives from elsewhere. Of particular importance were various versions of political economy, particularly those informed by Marxism and by world-systems analysis.[16]

In the early 1990s, while these perspectives remain influential, they have been supplemented, and in some cases challenged, by ideas drawn from social, cultural and (occasionally) political theory. A new wave in the development of political geography is emerging. This book charts its beginnings.

Notes to Chapter 1

1 Geoff Mulgan, *Politics in an Antipolitical Age* (Cambridge, Polity Press, 1994), pp. 7–8.
2 Relationships which are 'stretched' across time and space in this way are subject to what Anthony Giddens calls 'time–space distanciation'. See Anthony Giddens, *The Constitution of Society* (Cambridge, Polity Press, 1984).
3 I discuss these circumstances in more detail in Chapter 3.
4 The literature of the so-called 'new' cultural geography is rich and varied. For an introduction see: Kay Anderson and Fay Gale, eds, *Cultural Geography: Ways of Seeing* (Melbourne, Longman Cheshire, 1992); Denis Cosgrove and Stephen

Daniels, eds, *The Iconography of Landscape* (Cambridge, Cambridge University Press, 1988); Peter Jackson, *Maps of Meaning* (London, Unwin Hyman, 1989); Linda McDowell, 'The Transformation of Cultural Geography', in Derek Gregory, Ron Martin and Graham Smith, eds, *Human Geography: Society, Space and Social Science* (Basingstoke, Macmillan, 1994), pp. 146–73; Chris Philo, compiler, *New Words, New Worlds: Reconceptualising Social and Cultural Geography* (Lampeter, Social and Cultural Geography Study Group, 1991).

5 Of course, precisely because social processes are both discursive and material, it is difficult to separate the ideas of 'mode of production' and 'discursive formation'. Modes of production are themselves produced in part through discourses (such as those associated with property relations, for example), while discursive formations are produced materially and have material preconditions and consequences (for example they are dependent on the material means of information circulation). The point of distinguishing them here is to ensure that both aspects are held in mind.

6 Thanks to Miles Ogborn for this point.

7 Michèle Barrett, *The Politics of Truth: From Marx to Foucault* (Cambridge, Polity Press, 1991), pp. 129–30.

8 I owe the term 'mutual constitution' to Philip Crang.

9 Although it is not in any rigid sense 'structuationist', this framework clearly has some affinities with the work of Anthony Giddens, and I have found many of Giddens's ideas helpful. See Giddens, *Constitution*. One constraint, from my perspective, is that Giddens's stress on the routinized aspects of human agency limits his ability to deal with purposeful strategic action directed towards social change; with political practices in other words.

10 Giddens, *Constitution*, p. 35.

11 See, for example: Doreen Massey, 'New Directions in Space', in Derek Gregory and John Urry, eds, *Social Relations and Spatial Structures* (Basingstoke, Macmillan, 1985), pp. 9–19; Ed Soja, 'The Socio-spatial Dialectic', *Annals of the Association of American Geographers* 70 (1980), pp. 207–25; Ed Soja, *Postmodern Geographies: The Reassertion of Space in Critical Social Theory* (London, Verso, 1989).

12 Michael Dear, 'The Postmodern Challenge: Reconstructing Human Geography', *Transactions of the Institute of British Geographers* 13 (1988), pp. 262–74.

13 John Short, *An Introduction to Political Geography* (London, Routledge, 2nd edn 1993); Martin Glassner, *Political Geography* (New York, John Wiley and Sons, 1993).

14 I should immediately add that although political geography is one of the most male-dominated of all the sub-disciplines, women political geographers are working hard to change this state of affairs, as recent issues of the journal *Political Geography* demonstrate.

15 Glassner, *Political Geography*, p. 280. Emphasis added.

16 See, for example, Ron Johnston, *Geography and the State* (Basingstoke, Macmillan, 1982); Peter Taylor, *Political Geography: World-economy, Nation-state and Locality* (London, Longman, 1985; 3rd edn 1993).

|2|

The geography of
state formation

Overview

In this chapter I consider the historical geographies which gave rise to the
global system of modern states which we know today. I start by outlining
some of the ways in which geographers have sought to understand the state
in the past. I then discuss the relationships between space, place and the
global jigsaw of modern states. The next sections of the chapter consider
some of the main conceptual issues surrounding the state: the process of
state formation, the relationship between 'high' and 'low' politics and the
notion of sovereignty. I then move to looking in more detail at the processes
which gave rise to the typical form of the territorial state. Two processes are
emphasized: the preparation for and the waging of wars on the one hand,
and the building of the administrative systems of the state apparatus on the
other. I then conclude by looking beyond Europe to consider some of the
reasons why a political form which developed in one small part of the world
has become the dominant system of territorial organization throughout the
world.

States in space

A recurring theme

The political geography of states has been one of the longest running themes
in the sub-discipline of political geography. One of the people widely cred-

ited with founding the subject, Friedrich Ratzel, wrote extensively on the subject of the state. For Ratzel, the state was an organic, living entity, consisting of a relationship between a people, their culture and their territory:

> Some number of people are joined to the area of the state. These live on its soil, draw their sustenance from it, and are otherwise attached to it by spiritual relationships. Together with this piece of earth they form the state. For political geography each people, located on its essentially fixed area, represents a *living body* which has extended itself over a part of the earth and has differentiated itself either from other bodies which have similarly expanded by boundaries or by empty space.[1]

Ratzel's ideas cannot be understood outside the political and geographical context in which he was working. Writing towards the end of the nineteenth century, Ratzel was especially concerned about the political position of Germany and the German state in a Europe which was in the midst of constructing huge overseas empires (see Chapter 4). His concept of *Lebensraum* (living space) emphasized the connection between what he saw as the cultural superiority and vibrancy of the German 'nation' and the geographical territory to which it was 'constrained' in the middle of Europe. In stressing the organic connection between a nation and its culture and territory, Ratzel was continuing a discourse which began with the writings of the German idealist philosopher G. W. F. Hegel some eighty years earlier. Hegel saw the state as an 'idea' or the expression of the 'mind' of a people. Ratzel was responsible for spelling out what he saw as the territorial or geographical implications of the state.

In the mid-twentieth century, the American geographer Richard Hartshorne allocated the state a special role in his attempt to outline the *Nature of Geography* in 1939. In drawing up a case for the distinctiveness of geography as an academic discipline, Hartshorne sought to provide firm definitions of its constituent parts. The state, he declared, was the defining subject-matter of the sub-discipline of political geography. However, he tried to distance his conception of the state from that of Ratzel and his successors. This was partly because in the 1930s, the Nazi Party under Adolf Hitler had used Ratzel's term *Lebensraum* to provide scientific legitimacy for German territorial claims during the Third Reich. Hartshorne's notion of the state, therefore, was shorn of many of its explicit political associations and made into the expression of territorial administration. In his broad view of geography as 'areal differentiation' Hartshorne's concept of the state referred to the differentiation of territory into political units behind recognized boundaries.

As a territorial form, therefore, the state is the basic building block of the world political map. Traditional political geography has emphasized the geographical form of states in absolute spatial terms – their borders, land areas, and even shapes.[2] It has also been interested in the forces which promote or disturb territorial integration within states ('centripetal' and 'centrifugal' forces respectively) and territorial differentiation of states from their neighbours. This led to a focus on issues such as the role of transport and communications networks.[3]

With the emergence since the 1970s of approaches to human geography grounded in various forms of critical theory, geographers' understandings of the state have taken new paths. The dominant critical perspective during the late 1970s and early 1980s was informed by Marxism, and in political geography this led to an interest in Marxist writings on the state. Paradoxically, Marx himself had relatively little to say on the subject, but the role of the state (particularly in relation to capitalism) has been the subject of much debate among twentieth-century Marxists and other writers on political economy. These debates provided the material for a reappraisal by geographers of previous geographical writing on the state, and the early 1980s saw two books on the state by geographers, both of which drew heavily on forms of Marxist analysis.[4]

Marxism certainly provided state theorists with some powerful conceptual tools with which to beef up the rather descriptive approach of traditional political geography. It did this especially by showing the extent to which state policies, state élites and state finance were bound into the social relations of capitalism and the processes of capital accumulation. Many Marxist accounts, however, based their explanations of the state on the operation of economic processes. One difficulty with this is that it makes it difficult to interpret the variety of state forms, activities and histories among countries with, broadly speaking, the same economic system – capitalism. In addition, relatively little attention was paid to states which existed on the periphery of the capitalist system, or, as in the case of state socialist societies, largely outside it.

In other words, it became clear that it would be difficult, if not impossible, to produce a general theory of 'the' state, which would apply to all examples. As is often the case, however, academic enquiry has been overtaken by political events, and human geographers and other social scientists are being forced to re-evaluate their previous ideas about the relationships between states and societies. Those political events include: a new phase of state building, with the fragmentation of eastern Europe and the former Soviet Union into a much larger number of political units; a growing debate about state capacity both in the West (with the apparently growing inability of governments to intervene effectively in the economy to deal with unemployment and economic recession) and the economically impoverished countries of the South[5]; and an increased interest in constitutional issues (for example, with the growth of supra-national organizations). After some decades when the territorial state based on liberal democracy (see Chapter 3) was widely seen, at least in the West, as both 'normal' and desirable, questions about the forms, functions and even existence of such states are back on the agenda.

Perhaps 'new times' do not always require new theories to match, but they should at least prompt us to evaluate the theories we have been using to see whether they still pass muster. In this light, geographers working on states and their changing roles have increasingly turned to social, political and cultural theory, and to concrete accounts produced by historical sociologists. A range of questions and issues have been newly highlighted (or sometimes revisited). They include questions of war, militarism and violence; of bureaucracy, organization and surveillance; of culture, discourse and meaning; and of authority, citizenship, rights and resistance. Despite the questions that have been raised over its head, the state remains, for the moment, what the sociologist Anthony Giddens calls the most important 'power container' of modernity.[6] I have extended my discussion of this over two chapters. In this chapter I consider the geography of state formation. In the following one I look at the crisis tendencies in modern welfare states and the geography of their contemporary restructuring.

The global jigsaw

I want to start my substantive discussion of politics and geography with thinking about states at some length because they represent the foremost claimant of authority in the modern world.[7] No other set of agencies asserts its power over us quite so insistently (some would say insidiously) as the states in which we live. I will come back to the issues of power and authority later once I have raised some problematic questions about how we define states. To start with, though, we can get quite a long way with an everyday understanding of states as territorial units. There is, however, rather more to the relationship between states and geography than surface area.

We are all, I imagine, used to the political map of the world in which the land surface is divided up, almost completely, into the territorial areas called states.[8] To be sure there are a few blurred edges, especially where wars have left territorial disputes unresolved. In the vast majority of cases, though, we reside in places which are each clearly within the territory of one particular state, with clear boundaries separating it from its neighbours.[9]

This situation seems so normal to us that it is difficult to imagine how things could be otherwise. The difficulty of thinking outside the framework of states is demonstrated in part by the problems in finding solutions to many territorial political conflicts. In Palestine, for example, two social groups have both laid claim to the same territory, and both insist on their right to establish a state. Since, in our normal way of thinking, no two states can occupy the same territory it is impossible to reconcile both demands simultaneously. In the modern world, achieving statehood has been made to seem the ultimate goal for any group defining itself as a nation. Yet there is nothing inevitable or natural about states. Like all human institutions they are products not of nature, but of social and political processes.

They are, moreover, extremely recent creations. Human beings emerged some 400,000 years ago, but it was not until 8,000 years ago that anything that might be called a state appeared.[10] Further, for most of the time since then, states of whatever form have only occupied a small part of the earth's surface. It is only in the last 300 years that distinctively modern states have developed and only in the last 50 years that the modern form of the state has become more or less universal.

Even today the variety of state forms is quite large, and for most of the modern period the characteristics of different states have been highly diverse. It is therefore not only difficult, but in many ways downright misleading to try to construct a theory of 'the' state. The problem of trying to come up with a definition of 'the' state is that it depends on identifying the essence of stateness, as it were. Because states are political and social institutions, they are in a continuous (albeit slow) process of change and mutation: if we define the essence of the state in one place or era, we are liable to find that in another time or space something which is also understood to be a state has different essential characteristics.

In order to avoid such 'essentialist' interpretations, we need to give due weight to historical and geographical differences in the nature of states. Essentialist accounts also often try to ground an interpretation of the state in a central unifying principle, such as Hegel's state idea, or, as in (some forms of) Marxism, the mode of production. By contrast, I want to suggest that states should be seen as both complex networks of relations among a (shifting) mixture of institutions and social groups, and the product of their own processes of institutional development and historical change as well as important external influences.

If states as we know them today are so recent and (historically) so unusual, why is it that we tend to think they represent the natural order of things? Despite their (relative) newness, they are the most powerful organizations in the world. As such they embody and are constituted from a huge range of political and social interests, and exercise power in often highly unequal ways. A discourse which constructs the modern system of territorial states as *natural* serves to promote many of the interests and power relations involved as similarly natural. 'Naturalizing discourses' are among the most important forms of discourse, because they make what is social (and therefore changeable) seem natural (and therefore eternal). Although the close relationship between a state and its territory is central to this discourse, it is only one way in which geography enters into the constitution of states.

Geography and states

Let us focus for now on modern states, and retain, just for the moment, an everyday definition of the state as an organization with *de facto* responsibility for the government and administration of a territory. That territoriality is clearly a central feature of the geography of states, but geography is important in other ways too.

First, the territories of modern states are ordered by relatively precise boundaries. This is a largely taken-for-granted feature of the modern world. Of course, the positions of the boundaries may often be contested and, despite the best efforts of many traditional political geographers, it is fairly clear that there are rarely any purely technical or rationalist solutions to such conflicts. The existence of such disputes does not, however, undermine the principle that modern states are bordered by clearly demarcated linear boundaries: indeed, it strengthens it. Like the existence of states themselves, however, there is nothing natural about precise boundaries. I have already mentioned the work of the sociologist Anthony Giddens, and, as we shall see, his ideas are particularly helpful in thinking about the character and role of states. In Giddens's view, precise borders only emerge with modern states, and are associated with the capacity of states to spread their power relatively evenly throughout a territory. In earlier times, because of technical, resource and organizational limitations, state power tended to be much stronger in the centre of the territory than towards its edges. States had frontier zones, rather than borders, where the weak influence of one or more states overlapped, or state power petered out into areas not occupied by any state.[11] The neat boundaries of modern states, therefore, are symptoms of the ability of those organizations to project their administrative capacity across the whole of their territories.

Second, most modern states occupy large territories, and seek to administer them through various systems of territorially fragmented institutions. These range from loose confederations at one end of the scale, through federal systems (such as those in Germany or the United States) and systems of regional and local government (city councils, for example), to local offices of the central state at the other end (such as a local tax office).

Third, geography is important in the spatial structures of state institutions. An important (albeit difficult) set of ideas about the state are developed in the writings of Nicos Poulantzas. Poulantzas refers to the 'institutional materiality' of the state: to the material presence of the organizations and institutions of the state 'apparatus'.[12] These institutions obviously have spatial structures which take physical forms – offices, courts, parliaments, military bases and so on must be located somewhere (and where they are located can make a difference to how they work and what effects they have). In addition, though, it may be useful to think of them as having spatial structures which are social and symbolic too. Parliament buildings, for example, embody certain meanings. They form part of various discourses (about state power, for instance, or 'democracy'). They are also physically exclusive, setting and policing limits about who is entitled to speak politically or to govern (even in systems constructed, discursively, as 'democratic').

Fourth, and related to Poulantzas' argument, the apparatus of the state, spread throughout the territory, provides the means by which the state monitors, governs and attempts to control the population. I shall have more to say about surveillance shortly, but the capacity to keep an eye on what is going on clearly depends both on the territorial reach of the state, and on what we might call the spatial density of the mechanisms and practices through which monitoring occurs. These include the physical surveillance of space by police and other state employees and, increasingly, electronic surveillance using cameras. Less obviously they also include technologies of record-keeping and data-gathering, through which the activities of the population are monitored either at the aggregate level (through the collection of statistics[13]) or the individual level, through personal records relating to birth, marriage, death and a whole range of other aspects of our lives. New technologies are important here too, with the growth of smart-card electronics raising concerns about the state's potential future use of identity documents.

Fifth, according to Giddens, this monitoring activity has a tendency to increase over time in modern states. However, it is never absolute (even in societies labelled 'totalitarian'). This means that there are always gaps in the state apparatus in which resistance of various forms may develop – spaces of resistance, if you like. In the former Soviet Union there were networks of dissidents in which ideas and literature officially banned by the state were able to circulate (albeit in a highly restricted way), and in many countries state authorities either tolerate popular protest over a whole range of issues or do not have the resources to prevent it.

Finally, geography as *place* is significant in the composition of states. For example, at particular times, dominant groups may pursue deliberate state-building strategies (perhaps after independence from a colonizing state). In these circumstances a discourse of the state as 'homeland' is likely to develop as a means of legitimating the state.

I will discuss a number of these points in further detail below. First, however, we must consider the process of state formation in a bit more depth.

States and state formation

Problems of definition

There is no universally accepted definition of 'the state'. In part this is because, as with all objects of social scientific investigation, different writers adopt different perspectives and understandings, and these inform their definitions. In part though, the problems of definition arise from the difficulties of essentialism, which I mentioned above. If states are historically changing and their forms, functions, and meanings are open to conflict and contestation, what the state is and does can be very different in different times and places.

As we have seen, although it is often the subject of naturalizing discourses, the state is not inevitable, nor are particular forms of the state. Certain political theories adopt *normative* approaches to the state. That is, they try to spell out what the state should be like and what a 'good' state would be. Such theories are sometimes linked to notions of progress, usually towards 'democracy'. This has led some Western governments, for example, to argue that Western-style 'liberal democracy' is the best type of state and should be universally adopted. In practice such theories are not the detached and objective arguments they sometimes claim, but are part of the process by which governments and states claim legitimacy. By naturalizing the liberal democratic state they make what is a highly unusual and geographically specific state form seem 'normal' – the culmination of human progress to date. This is related to what is known as the 'Whig interpretation of history' – the argument that human history has been leading up to the present day and that past forms of social and political organization should be evaluated according to how far they advanced or retarded that process of development.

I want to adopt a rather different approach. In my view it is important to recognize that the so-called liberal democratic state is one quite particular state form. Not only is it not 'natural' or 'normal' in any absolute sense, but it is also the product of specific historical, cultural and intellectual circumstances (see Chapter 3). This means that any definition of the state must give due weight to other types of state, including those which for various reasons faded away, and to the dynamic, complex and shifting character of all states.

By way of a working definition, therefore, I propose the following:

> States are constituted of spatialized social practices which are to a greater or lesser extent institutionalized (in a 'state apparatus') and which involve claims to authority which are general in social scope and which secure at least partial compliance through either consent, or coercion, or both.

This may seem disappointingly vague, but there can be no detailed definition of the state which is both transhistorical (valid throughout time) and applicable to the wide variety of social forms which have been understood to be states in different historical and geographical settings. The suggestion that the claim to authority should be general distinguishes states from, for example, the management of a firm. The argument that the claim should actually be complied with (at least to some extent) ensures that the term 'state' is only applied to institutions which are recognized as such (at least by some) and which are therefore effective, and not to fringe organizations which might claim authority, but which have no means of making the claim stick. The definition proposed above could include groups which are not conventionally regarded as states (such as some religious organizations). In my view, however, in so far as some of their practices fit the definition, then they could in fact be viewed as more or less state-like. This open definition thus ensures that account can be taken of the emergence of new types of states, or state-ish practices. Thus the United Nations, the European Union, monastic orders and the Mafia, among others, are all like states in some respects. Although none of them claim statehood, they all claim more or less general social authority in particular places and in some contexts effectively replace states as, for example, rule-makers and keepers, welfare providers and conflict-resolvers. We might refer to them as 'quasi-states'.

State formation as a social process

States, then, are not natural or inevitable, but are the products of specific social processes and political struggles which generate a process of state formation. However, it seems highly unlikely that a ruler or other dominant group could have coolly decided to deliberately create the modern form of the state before it had emerged. Even if such rational calculations had been undertaken it is even less likely that a far-sighted potentate would have had the technical and organizational capacity and all the resources required.

Certainly states are the products of the actions of people, both of the powerful and of the relatively powerless, whose labour and loyalty (or resistance and dissent) are part of the process too. However, in late medieval and early modern Europe, where modern states began to emerge for the first time, none of the people involved could have foreseen (let alone desired) the vast bureaucratic powerhouses which we recognize as states today. This means that the process of state formation was a by-product of other activities, which may in themselves have been intentional, but which were not intended to generate the multifunctional modern state. In the language of Anthony Giddens, modern state formation was the 'unintended consequence' of intentional activities.[14] However, Giddens also argues that human activities are 'reflexively monitored'.[15] What he means by this is that both individually and institutionally we continually examine our actions and their consequences. This means that once state institutions start to emerge, once they gain their 'institutional materiality', they become objects of human understanding and reasoning. This means that individuals and groups, both within and outside the state apparatus, start to pursue strategies in relation to the state. In some cases this may include attempting to emulate the formation of states elsewhere. This occurred, for example, in a number of former British colonies, where the British state was used, following independence, as a model for institutions of government and administration. Such strategies never arise on a blank surface, however; there is always a historical legacy – a set of institutions and conventions inherited from the immediate past. These provide the resources with which actors pursue strategies for the future, but they also limit the range of options. On a day-to-day basis, change is often piecemeal rather than dramatic, although piecemeal changes can add up to complete transformations over a long period. Occasionally, in the cases of revolutions, previous structures may be almost wholly dispensed with, although even here it is likely that the revolutionary strategy itself will have been heavily conditioned by the previous forms which were the context for its development. For example, historians studying Russia have often remarked on the extent to which the Tsarist state influenced that established by the Bolsheviks, following the Russian revolution in 1917.

Political strategies are important to state formation, therefore, but they rarely turn out as expected. They are also multifarious. The development paths of modern states have not been unilinear, even where similar forms have emerged in the end, and often the products have been markedly different. State formation should certainly not be seen as a process which tended automatically towards the modern form of the state. Along the road many other forms emerged, grew and declined: city states, absolutist monarchies, empires, satellites, religious governments and others rose and then, for the most part, fell. Their falls should not be seen as indicating that the surviving modern form of the state is superior, either morally or functionally. Decline can, after all, occur for a variety of reasons, and may have happened in some cases in spite of the otherwise functionally and morally 'good' features.

State formation, therefore, is not a process in which a 'more effective', 'more democratic' or 'more enlightened' system of political administration arose from 'inefficient', 'despotic' or 'ignorant' predecessors. While medieval states were no doubt all of these things from time to time, it would be a mistake to assume that modern states have done away with all forms of domination, inefficiency and irrationality, as we shall see below.

High and low politics

State security and social security

Before outlining some actual examples of state formation it will be helpful to add to the concepts of formal and informal politics, which we considered in Chapter 1. The terms 'high politics' and 'low politics' might seem similar at first to the ideas of formal and informal politics, but they actually focus attention on a rather different division in the political process.

High politics refers to the politics of war, peace, diplomacy, the state's claim to sovereignty and constitutional change. It touches on the very existence of the state, and the ways in which it deals with threats to that existence. Its strategies commonly (though not exclusively) involve the people who occupy élite positions in the state apparatus. Since it is involved with the big questions of the state's existence and broad organization, we might think of it as dealing with state security.

By contrast, *low politics* refers to more mundane issues such as economic policy, public health, education, routine administration, welfare benefits and environmental protection: the kinds of issues over which states rarely, if ever, go to war, but which today occupy a large part of their attention and resources. The strategies and practices concerned do involve state élites, especially in producing legislation, but are carried out overwhelmingly by the ordinary personnel of the state – the junior civil servants, employees of municipal councils, teachers and social workers.

As in the case of formal and informal politics there is a degree of overlap between the two – they are not mutually exclusive. Thus the provision of social welfare may be used to assert the legitimacy of a state's claim to authority, while the protection of state security commonly involves the routine monitoring of many more ordinary people than is often realized. However, the high/low distinction is different from the formal/informal one. High politics involves informal politics (such as personal relations between heads of state) as well as formal politics (in the shape of diplomatic missions, constitutional commissions and the like). Similarly, low politics involves the formal arenas of parliament and civil service, as well as informal politics within a council department or an educational institution.

A shifting balance

The balance between high and low politics has moved back and forth over time. During war, for example, high politics comes to the fore. Over a relatively long period, however, there has been a tendency for high politics to decline in importance relative to low politics, at least in the West. In premodern states, government was dominated by high politics. Rulers were concerned first and foremost with issues such as territorial conquest and expansion, securing the constitutional succession for monarchic dynasties and gaining wealth and prestige relative to other states. The daily lives of their subjects were of very little concern to them, at least by contrast with the situation today. Provided the masses did not pose a threat to the state they were, for the most part, ignored. With the emergence of the modern form of the state, in which the state becomes distinct from the person of the monarch, the balance began to shift. More and more states became concerned with the everyday affairs of their resident populations. To begin with, this was a by-product of the state's increased demand for resources with which to finance its own activities. Raising more taxes required more knowledge and information on the population and its activities, and this led to a growing tendency to keep tabs on what was going on 'at home'. At the same time, doing more to and for ordinary people required its own kinds of resources. These were not just a question of money, but also required new forms of technology – such as the means to collect and record data.

The roots of the shift from high to low politics correspond to the development of new discursive formations. You will recall that the concept of discursive formation originated with the work of Michel Foucault (see Chapter 1). Foucault's early work was not much concerned with the processes of formal politics; indeed, it was instrumental in focusing much-needed attention on informal politics. In his later writings, however, he did consider the issues of state and government in more detail.[16] According to Foucault, the idea of government was not originally associated with what we now consider to be 'politics'. Initially 'government' referred to government of oneself – the exercise of self-control. The term then came to refer to government of the family or the household (by, for example, the father). Until the sixteenth century, Foucault suggests, the ruler of a state was concerned with preservation of the state, rather than with governing: 'to be able to retain one's principality is not at all the same as possessing the art of government.'[17]

The notion of government in its modern, political sense only arises when the management of the state comes to be understood in the same way as a father's management of a family. That is, that the governor (father) comes to be concerned with the ordering of the people, activities and things of the state (household) and with their interrelations. This is a different concern from simply ensuring the survival of the state or protecting the monarchy from overthrow. With the shift to what Foucault calls governmentality, the ruler of a state begins to take an interest in, and to pursue strategies towards, the people who live in the territory of the state, and their affairs, including economic activities, social norms and so on. Previously what the people did was of little concern to the prince unless they threatened the state. Central to this change was the identification of the people of the state as a *population* which was understood as the proper focus of the art of government. For Foucault, the discourses and practices of governmentality emerge during the sixteenth century together with the objects of government: the population of a particular territory.

A relative shift towards low politics involved increasing what I have called the density of relations between state and society. In order to provide public health measures, mass education, environmental improvements and welfare benefits, the state has to 'penetrate'[18] society much more intensively, which requires additional resources of all kinds: staff, institutions, buildings, knowledge, systems of organization and frameworks of understanding. While states are in general more highly militarized than ever, there has been a relative shift away from high politics with the decline of war-making[19] as a routine activity, at least in the West. Wars continue to be fought from time to time, but the routine role of states, on which the majority of state revenues are spent, has become dominated by low politics.

This shift is less marked in the impoverished South, where a much higher proportion of state resources and activities are commonly devoted to high politics. This is in part a consequence of the smaller overall resources available to Southern states. Since state security (both material and symbolic) is widely regarded (by state élites) as the first priority, it is common for the first call on resources to be allocated to military and diplomatic activities. Where resources are limited, this may leave little for anything else. The absence of successful economic policies and the lack of social welfare provision, may, of course, exacerbate the threat to the survival of the government, or even of the state, from 'below'.

A further implication of the relative lack of importance accorded to low politics prior to the full development of modern states was that government often did relatively little by way of governing. The notion of governance, as a routine, continuous and fairly intensive monitoring, regulation and administration of a wide range of activities in society does not arise where high politics is the order of the day, since this commonly involves the state in strategies which are either directed externally (to allies and enemies) or confined to the state élite (such as the court in absolutist monarchies).

A growing concern with low politics is partly the product of strategies from 'above'. Tax gathering, for example, both serves the immediate purposes of the state and involves low politics. More often, however, it is the result of pressures, or responses to perceived threats, from 'below'. The long-term trend towards low politics developed especially strongly in the context of the dramatic industrialization and urbanization of the nineteenth century. These changes (see Table 2.1) produced large, impoverished urban populations, removed from many of the traditional ties of rural life and less able to rely on local sources of support. Living and working conditions were often dangerous and unhealthy. Such dramatic transformations gave rise to popular social movements (see Chapter 6) which pressed for reforms and the provision of social welfare, health services and education. At the same time, regardless of the actual conditions in large cities, the urban poor were regarded by the wealthy and by the state as a source of disease, moral laxity and social unrest. The strategies of the poor and the fears of the rich constituted a pressure on the state for social reform which had not previously arisen. In addition, industrialization gave rise to a new set of social interests, of industrialists, capitalists and entrepreneurs, who were concerned that the state should turn its attention to economic matters and to trade policy. In the twin processes of industrialization and urbanization, therefore, lay the seeds of the two

Table 2.1 Urbanization in Europe west of Russia

Year	People living in cities of 10,000 or more (millions)	Percentage of population living in cities of 10,000 or more
1590	5.9	7.6
1790	12.2	10.0
1890	66.9	29.0
1980	c.250	c.55

After Tilly, 1990[10]

major concerns of twentieth-century low politics in industrial economies: economic progress and social welfare.

Although the two are linked, in this chapter I am more concerned with high politics – with state formation and the development of state institutions. In the next chapter I will give more attention to low politics in discussing the rise and apparent decline of welfare states.

Claiming sovereignty

No higher authority?

In the modern world, states are the foremost claimers of authority, an authority which is simultaneously claimed to be legitimate. In other words, states claim to have the right to require residents of their territories to behave in certain ways and to refrain from certain activities, that is, to receive compliance. The fact that these rights of the state are (within limits) more or less universally accepted in most states should not mislead us into thinking that they are absolute or can be legitimated in any permanent way. They remain claims and assertions, albeit ones which are conventionally accepted, both by residents and by other states. The discourse of sovereignty raises the stakes still further. As Joseph Camilleri and Jim Falk argue:

> Sovereignty is a notion which, perhaps more than any other, has come to dominate our understanding of national and international life. Its history parallels the evolution of the modern state. More particularly, it reflects the evolving relationship between state and civil society, between political authority and community. ... despite loose talk about the way it is acquired, lost or eroded, sovereignty is not a fact. Rather it is a concept or a claim about the way political power is or should be exercised.[20]

A claim to sovereignty is a claim to being the highest authority within an area, or over a particular group. Modern states' claims to sovereignty are conventionally recognized by other states, although some states are not regarded as sovereign or legitimate by all others. For example, the state of Northern Cyprus, is not recognized as legitimate by any European countries apart from Turkey, although it operates in most other regards as any other state. An institution would be entirely sovereign if there were no organization or institution which could require its compliance in any field of activity and if it were free to pursue its own policies unhindered, at least within its own territory. It is doubtful whether states have ever been sovereign in this sense. Initially, as they developed, other claimants to authority, such as kinship groups and religions, have often been able to hold sway, at least in some areas of life. In the contemporary world, the gaps in states' sovereignty are numerous. In some cases there are *de jure* competitors, such as the European Union, which has the capacity to legislate on a wide range of trade, employment and economic matters. There are also many *de facto* challengers, such as transnational corporations, and international monetary and aid organizations. For example, transnational corporations can effectively bypass certain taxation regulations by manipulating the prices charged by one arm of the corporation to another. Since the resources available to the largest transnational corporations rival those of many states, and because they are able to move those resources across international boundaries with increasing ease, there are many areas of economic policy-making in supposedly sovereign states which are certainly heavily influenced, if not actually determined, by the strategies of transnational corporations.

Origins of modern claims to sovereignty

Despite contemporary challenges to their claims to sovereignty, states remain the most powerful organizations on the planet, and have the resources to pursue a variety of strategies in support of their claims. The doctrine and discourse of sovereignty, however, developed in Europe in tandem with the modern state itself. In feudal Europe, power was in many ways highly decentralized. The broad normative and legal framework was regarded as fixed, and divinely ordained, rather than the product or possession of the government. The monarch may have ruled by divine right, and been regarded as the ultimate temporal authority, but was almost wholly detached from the daily lives of ordinary people. Power was exercised through a highly hierarchical, but simultaneously decentralized system. The local lord was a far more important (and powerful) figure in the everyday lives of villagers than their king or queen.

As the feudal system began to disintegrate, the power of monarchs was strengthened. The system of absolutist monarchies, which came to dominate Europe in the sixteenth and seventeenth centuries, saw the concentration of power (in principle, absolute power) in the hands of the monarch. According to Anthony Giddens, however, unlike medieval monarchs whose power was embodied in their very person, the sovereignty of absolute monarchs was in principle at least, separable from the individual known as the sovereign. This allowed a shift, with the growth of a centralized state apparatus which extended beyond the court, from the sovereign-as-monarch to the more impersonal 'sovereign state'.[21]

Of crucial importance to this process was the emergence of a number of absolutist states together and the resulting development of the interstate system. Since sovereignty cannot be grounded in any absolute foundations, it is constructed in practice through a system of mutual recognition. A state's claim to sovereignty is 'made to stick', as it were, by showing that other states regard it as a legitimate claim. In one sense this is a circular exercise, since those acknowledging the claim have an interest in getting their own claims recognized. None the less it has provided the preconditions for the hugely powerful system of territorial administration of modern states, by providing a bounded space in which the massive apparatus and complex practices of the modern state could develop relatively free, under 'normal' circumstances, from external intervention.

One of the key moments in this mutual recognition process came at the end of the Thirty Years War in 1648. The Treaty of Westphalia secured the foundations of the modern state system by agreeing that individual territorial states, rather than the empires of which they were a part, should have the right to conduct their own diplomatic relations with other states, and that in principle, states should be regarded by other states as sovereign within their own borders. Thus it can be seen that far from being a natural and universal norm, the modern territorial sovereign state is the product of quite particular historical circumstances. Moreover, notwithstanding the undoubtedly large resources of the modern state, its claim to sovereignty remains open to challenge and continues to be contested, albeit often implicitly, as with the growth of transnational corporations.

Rulers, resources and wars

High politics and state security

While episodes such as the Treaty of Westphalia are important, it is crucial to remember that state formation is an ongoing process. The process is not necessarily smooth, and indeed, Anthony Giddens argues for a 'discontinuist' view of history, which does not start from the assumption that present circumstances represent the neat unfolding of an even process of progressive development.[22] Things are certainly much messier than that. None the less, state institutions do endure over long periods, even if their functions may change (sometimes quite sharply).

Most states set a very high priority on securing their own survival. State security is usually presented by governments as their first duty. Notwithstanding the rise of low politics, modern states continue to this day to advance a variety of discourses concerned with protection from external and internal threats. In all cases, therefore, relations between states are, to a greater or lesser extent, discursively constructed. The implications of this insight for contemporary international relations are considered in more detail in Chapter 5. At the same time, one effect of these geopolitical discourses is to promote preparations for war. Such preparations still continue with the development, trade and stockpiling of weapons of all sorts all over the world. Modern states are typically highly militarized organizations, although the extent to which militarism is present in everyday life varies considerably.

State formation as the product of war

Charles Tilly argues that preparations for war and the waging of wars were crucial to the process through which the modern European state system developed, and that this has often been underestimated. He further argues that the ability of states to prepare for and wage war was heavily dependent on the kinds of resources available to rulers (he calls such resources 'capital'), and that rulers' strategies in relation to war generated state institutions and practices largely unintentionally. Moreover, the strategies which were pursued, and thus the processes of state formation, were strongly affected by the strategies and institutions of other rulers and states.[23] His analysis therefore fits very closely the approach to politics which I outlined in Chapter 1.

Preparing for, and waging, war are influential in state formation for a number of reasons. Making war is expensive and complicated. It requires large resources of people and equipment and significant levels of organization. It therefore requires taxation, recruitment into the armed forces and the development of new institutions. According to Tilly, these form the core of the process of state formation. In addition 'successful' wars may increase the territorial possessions of rulers, and start to demarcate the more precise boundaries associated with modern states. Of course this process relies heavily on coercion rather than consent:

Why did wars occur at all? The central, tragic fact is simple: coercion *works*; those who apply substantial force to their fellows get compliance, and from that compliance draw the multiple advantages of money, goods, deference, access to pleasures denied to less powerful people. Europeans followed a standard war-provoking logic: everyone who controlled substantial coercive means tried to maintain a secure area within which he [sic] could enjoy the returns from coercion, plus a fortified buffer zone, possibly run at a loss, to protect the secure area. Police or their equivalent deployed force in the secure area, while armies patrolled the buffer zone and ventured outside it; the most aggressive princes, such as Louis XIV, shrank the buffer zone to a thin but heavily-armed frontier, while their weaker or more pacific neighbors relied on larger buffers and waterways. When that operation succeeded for a while, the buffer zone turned into a secure area, which encouraged the wielder of coercion to acquire a new buffer zone surrounding the old. So long as adjacent powers were pursuing the same logic, war resulted.[24]

The extent of war during the period of the emergence of modern states was dramatic (see Table 2.2).

Wars are expensive, requiring the maintenance of large armies who are not engaged in production, and the acquisition of equipment, much of which has to be continuously replenished. Paying for all this involves taxation (in a broad sense), or borrowing against future taxation. Certain forms of taxation such as direct and arbitrary collection of money or goods (which Tilly calls 'tribute'[25]) can be garnered by *ad hoc* and coercive means. More systematic and dependable taxation regimes require organization and monitoring of the population, and a monetary economy.

In addition, the armed forces had to be organized and managed, and as warfare itself became increasingly large scale and complex, so too did military organizations. According to Christopher Dandekar, the sixteenth and early seventeenth centuries saw a revolution in military organization:

Table 2.2   Temporal extent of war between great powers

Period	Number of wars	Proportion of period during which war was under way (%)
16th century	34	95
17th century	29	94
18th century	17	78
19th century	20	40
1901–75	15	53

After Tilly, 1990[10]

The state transformed military organization from a system comprising autonomous, largely self-equipped mercenary formations, employed by contracting captains, to one based on professional servants of the state, disciplined in a bureaucratic hierarchy and owing allegiance to the state alone.[26]

Despite the importance of coercion, the dependence of the warmongering states of early modern Europe on the wider economy and society for the resources for war, forced their ruling elites into relations of strategic alliance and compromise with other social groups. These other groups were often pursuing very different interests and strategies, but they were sometimes able to secure their aims in a process of negotiation with the dominant élite:

> In fact, rulers attempted to avoid the establishment of institutions representing groups outside their own class, and sometimes succeeded for considerable periods. In the long term, however, those institutions were the price and outcome of bargaining with different members of the subject population for the wherewithal of state activity, especially the means of war. Kings of England did not *want* a Parliament to form and assume ever-greater power; they conceded to barons, and then to clergy, gentry, and bourgeois, in the course of persuading them to raise the money for warfare.[27]

These processes and strategies resulted in the emergence of state institutions from which developed modern states. Among the key institutions involved were treasuries, state banks, taxation departments, diplomatic corps, military administration, military academies, armies and navies, and, as a product of bargaining with other social groups, (partially) representative institutions, such as parliaments. These organizations formed the cores of the apparatuses of modern states. On the whole none of them were established deliberately in order to construct modern states – the idea would have made little sense to those involved at the time. Rather they were the by-products, the unintended consequences of strategies pursued for other reasons, most notably the preparation for, and waging of, wars.

The character of states: variations, then convergence

During the Middle Ages the political map of Europe was both fragmented and complex. Hundreds of rulers governed a multifarious patchwork of statelets, cities, dukedoms, principalities, caliphates and larger empires. Within the largest units (such as the dynastic empires) dozens of local potentates pursued their own interests and strategies largely independently from those of their ultimate overlords.[28] As state formation proceeded through the pursuit of war, different states developed in very different ways. By the middle of the twentieth century, however, states had become much more alike. While significant differences remain, modern states have much more in common with each other in terms of their activities and forms of organization than did their early modern ancestors. This process of differentiation and then convergence is central to Charles Tilly's account of state formation.

The resourcing of armies and wars through various forms of taxation over a sustained period depended upon the ability of the economy to generate sufficient production to maintain not only the general population, but also the military activities of the state. According to Tilly, the capacity of states and their rulers to pursue militarist strategies was heavily influenced by the reciprocal relationship between coercion and capital. We have already seen the importance of coercion, but it was the various different ways in which coercion combined with the availability of capital that led to variation in the character of states.

The spatial structure of the relations between capital and coercion were of crucial importance. According to Tilly, the means of coercion were characteristically mobilized by states and their rulers. Capital, by contrast, was concentrated in cities: the home of banks, merchants, traders, markets and craft workers. Anthony Giddens argues that while states have become the pre-eminent power containers in the modern world, it was cities which held that position in earlier societies.[29] Cities were in many respects rivals to the emerging states. Cities had their own institutions and resources and were concerned above all with production and trade, rather than war and the acquisition of territory. In some cases, they formed states in their own right – city states. In others they existed more or less uncomfortably *within* the territories or spheres of influence of emerging states.

States and their rulers relied to a greater or lesser extent on the resources which cities could provide. The precise balance varied, and Tilly identifies three contrasting trends in state development: capital-intensive, coercion-intensive and an intermediate 'capitalized coercion' path. These alternatives were not deliberate strategies, but represented the response of states to the different environments in which they found themselves.

In the capital-intensive mode, 'rulers relied on compacts with capitalists – whose interests they served with care – to rent or purchase military force, and thereby warred without building vast permanent state structures.'[30] By contrast, where coercion dominated, 'rulers squeezed the means of war from their own populations and others they conquered, building massive structures of extraction in the process.'[31] The intermediate path involved aspects of each, and included 'incorporating capitalists and sources of capital directly into the structures' of the state.[32] According to Tilly, examples of the first approach include Genoa and the Dutch Republic, of the second Brandenburg and Russia, and of the third, France and England.

In due course, however, the loose federations of city states at one end of the scale and the massive tribute-taking empires at the other both lost out to the 'intermediate form', the modern state:

> Which sort of state prevailed in a given era and part of Europe varied greatly. Only late in the millennium did national states exercise clear superiority over city-states, empires, and other common European forms of state. Nevertheless, the increasing scale of war and the knitting together of the European state system through commercial, military, and diplomatic interaction eventually gave the war-making advantage to those states that could field standing armies; states having access to a combination of large rural populations, capitalists, and relatively commercialized economies won out. They set the terms of war, and their form of state became the predominant one in Europe. Eventually European states converged on that form: the national state.[33]

Tilly's arguments are important, and the differing relationships between capital and coercion in different states were certainly highly influential in shaping different state forms. However, there is another aspect to the process of state formation which was also important in the constitution of states and the differences between them. In their book *The Great Arch*, Philip Corrigan and Derek Sayer[34] argue that state formation should be seen as a *cultural* process. So far we have looked at the ways in which military strategy and the relation between coercion and capital shaped European states. It should be remembered, however, that these aspects, together with the institutions which they produced are cultural, as well as military, political and economic phenomena.

What does it mean to say that state formation is cultural? First, it implies that it is a process which is symbolic as well as organizational or material. State institutions and practices embody a wide range of meanings, in their buildings, spatial arrangements, discourses, flags, costumes, ceremonies and routine activities. A berobed and bewigged judge symbolizes something different from one wearing a business suit. The elaborate ceremonials of the British monarchy carry a different set of coded meanings from the more austere rituals of a federal republic such as the United States. A conscript army means something different from one made up of volunteers. A parliament which meets in an ancient and grand palace is governing (symbolically) in a different way from one which meets in a purpose-built modern office building. Second, it implies that the production of meaning is central to the progress of state development. The state is not only a set of institutions, but also a set of understandings – stories and narratives which the state tells about itself and which make it make sense (in particular ways) both to its personnel and to the general population. These might include myths and legends, the official history of the state, or fictions and dramas which represent the state, its people and government in particular (usually heroic!)

ways. Third, there is a sense in which state activities are *performed* by the actors involved.[35] State bureaucrats behave in bureaucratic ways because they have an understanding of what it is bureaucrats do, with which they try to fit in. Armies, police forces, tax inspectors, administrators, teachers and politicians all work with a set of cultural codes about what it is to be a soldier, police officer, tax inspector and so on.

Crucially, cultural aspects can vary markedly between different states as their formation progresses. Even modern state institutions like parliaments, or bureaucratic departments, which may seem organizationally similar may have very different effects and roles as a result, in part, of the different discourses, symbols and performances embedded within them. According to Corrigan and Sayer[34], it is these cultural differences, which account for much of the distinctiveness of the English state as it developed from the Middle Ages onwards. Among other things, they emphasize the discourses of the state and its role in moral regulation as key aspects of its cultural formation. The discourses of the state are multiple (and sometimes contradictory). They include legislation, court judgments, inquiries, regulations, official reports, histories, educational material, public pronouncements and political arguments. The work of Corrigan and Sayer suggests that over time, these discourses, through their rhetorics, characteristic language and symbolic content, serve to mould the state as a series of cultural forms.

For Corrigan and Sayer, the concept of 'moral regulation' carries a broad meaning – much more than legislating against 'vice', for example. By moral regulation, they mean the processes by which the state tries to represent itself as the neutral guardian and protector of a unified whole people, which is actually a heterogeneous mixture of different and often conflicting social groups and interests. The state tries to pull together and integrate society, in part by representing itself as the embodiment of society. How often, for example, do we hear journalists speak of the British or the Americans or the Chinese when what they actually mean is the British government or the US administration or the Chinese authorities? The widespread confusion of states with their populations is evidence of the success of state strategies in trying to represent themselves as normal and natural expressions of a homogeneous and united people.

This process is always contested, however, with more or less success. Distinctive social groups, such as nationalist minorities, political opponents of the state, social classes, and social movements of all sorts have been at pains to undermine this claim of the state to what we might call cultural authority. (This and other aspects of social movements are discussed in more detail in Chapter 6.) Such opposition to state authority rarely goes unanswered, and while an emphasis on war with other states has been an important part of my argument so far, I want to turn now to the question of the state's control of its internal dissenters and resident populations.

Administrative power and state apparatus

Power and information

Recent work by sociologists and historians has stressed the vital importance of surveillance, monitoring and internal control for the development of states. For Anthony Giddens, 'administrative power' is one of the defining features of modern states. Its origins date back to the development of writing, and the importance of recorded information and information storage to the emergence and power of traditional states. In pre-modern societies technological and resource limitations prevented the kinds of detailed information storage that we now associate with all kinds of large organizations, but especially states. None the less the fact that some kind of recording, however limited, was undertaken was a key breakthrough and gave rise to the very possibility of the state as an administrative organization.

Writing and information storage and retrieval allowed a gradual shift away from power as the immediate expression of the will of the monarch towards power as the capacity of institutions to co-ordinate large-scale resources for strategic objectives. The sociologist Michael Mann has drawn a useful distinction between despotic power and infrastructural power.[36] Despotic power refers to the power of state élites to do things without reference to the rest of society. As Mann graphically puts it: 'Great despotic power can be "measured" most vividly in the ability of all those Red Queens to shout "off with his head" and have their whim gratified without further ado – provided the person is at hand'.[37]

By contrast, infrastructural power refers to the ability of the state to 'penetrate' civil society and reach out across geographical space to influence events throughout its territory. States which were despotically strong, but infrastructurally weak, had great powers over life and death in theory, but did not possess the logistical wherewithal to carry them out. Where states have great infrastructural power, but limited despotic power, they typically have huge bureaucracies reaching into every part of the land, but are unable to use them to produce rapid or effective results. Mann identifies a range of types of states depending on the combination of infrastructural and despotic power (see Table 2.3).

Integral to infrastructural power is the collection of information. As we have seen, absolutist states financed their military activities through taxation, and this was one of the first contexts in which systematic record-keeping was undertaken. For Giddens, however,

As good a single index as any of the movement from the absolutist state to the nation-state is the initiation of the systematic collection of

Table 2.3   Despotic and infrastructural power

	Infrastructurally weak	Infrastructurally strong
Despotically weak	Feudal	Bureaucratic
Despotically strong	Imperial	Authoritarian

After Mann, 1988[36]

'official statistics'. In the period of absolutism, such data-gathering was particularly concentrated in two areas, at least as regards the internal affairs of states. One was that of finance and taxation, the other the keeping of population statistics – which tended, however, until the eighteenth century to be localized, rather than centralized. . . . The official statistics that all states began to keep from about the middle of the eighteenth century onwards maintain and extend these concerns. But they also range over many sectors of social life and, for the first time, are detailed, systematic and nearly complete. They include the centralized collation of materials registering births, marriages and deaths; statistics pertaining to residence, ethnic background and occupation; and . . . 'moral statistics', relating to suicide, delinquency, divorce and so on.[38]

Surveillance and pacification

According to Giddens, there are four axes which define the modern age. These are industrial production, capitalism as a way of organizing that production, heightened surveillance, and the centralized control of the means of violence. The state is involved to varying degrees in all of these, but is most fully involved with the last two: surveillance and pacification. Those of us who live in modern states with a high degree of infrastructural power would find it very difficult, probably impossible, to pursue our everyday lives entirely independently from the state. The state monitors our births, marriages and deaths, our work and income, our child rearing, our health, our housing, transport and travel, our entitlement to public assistance, our political activities, our law-breaking and much else besides. Some of this information is held anonymously, but much of it is in named records. With the technological changes associated with the development of microprocessors, electronic information storage has greatly expanded the ability of states to keep tabs on its population. On the whole information-gathering is *not* undertaken by specialized security services, although some of it undoubtedly is. Rather it is the by-product of a huge range of routine daily interactions between people and state institutions.

Geography is crucial to the state's capacity to undertake such routine surveillance. It requires a high level of infrastructural power, and thus depends on a spatially dense and comprehensive set of institutional practices through which whole populations, from Miami to Seattle or from Land's End to the Orkneys, can be drawn into the knowledge circuits of the state. The 'institutional materiality' of the state which I mentioned above has a geography which stretches its practices throughout the state's territory, usually, although not necessarily, through a spatially dispersed network of offices, courts, registries and agents. This allows the expansion of state power away from the centre and right up to the boundary, enabling the establishment of the kinds of sharply-drawn borders characteristic of modern states.

In addition to surveillance, administrative power is expressed through the process of what Giddens calls 'internal pacification'. In traditional states, the centre had very little capacity to suppress internal dissent or unusual behaviour. Giddens mentions two developments which led to an increased emphasis on the state suppression of what gradually came to be defined as 'deviance'.[39] First, the growth of a large class of landless and dispossessed people led to rural unrest, poverty and rapid urban growth. Second, the state became increasingly concerned with the separation and treatment of specific social groups constructed as 'deviant' or 'abnormal'. These included those suffering from mental distress (the 'insane'), those with certain diseases, those committing criminal offences and those regarded as immoral or morally degraded, such as prostitutes and unmarried mothers. These twin shifts produced further parts of the state apparatus: the police, the internal security forces and 'carceral' institutions, such as prisons, workhouses and mental institutions.[40]

Resisting the state

The surveillant state is not without challenges, however. In part it carries on its detailed surveillance precisely because challenges to it exist. But the citizens of modern states are not dupes. They deploy a whole range of forms of resistance to state power. In all countries of the South and the former Eastern bloc and in a growing number of Western countries a burgeoning informal economy has grown up, outside the monitoring activities of the state. In some cases the goods and services it provides have been obtained through arbitrary and violent means, such as burglary and interpersonal violence, which seem to echo in a small way the excesses of the absolutist monarchies. In other cases, however, community groups working co-operatively have developed alternative trading and banking systems (so-called LETS: Local Exchange and Trading Systems) which bypass the formal structures of state power, and which bring much-needed economic growth and employment to impoverished communities. Much of the increasing capacity of the state to undertake monitoring and information-gathering is a product of the development of electronics and the rise of new information and communication technologies (NICTs). On the other hand, the same technologies can be used to transfer information between groups and individuals opposing state authoritarianism. Such information can now be transferred quickly, quite cheaply and largely unseen across international frontiers and stored in very small spaces. Through international computer networks (the Internet) it can also be published and transmitted to an increasing number of people. It is too early to tell what impact popular use of NICTs will have on the ability of states to control their populations, but at the very least it is clear that technological development is not always inherently and uniformly in the interests of the very powerful.

Finally, no system of control is perfect or complete. As a number of writers have shown, there is always the possibility of some resistance. Sometimes this is symbolic; sometimes it takes the form of hidden acts of sabotage or non-co-operation. The strategies of state institutions are met with the tactics of everyday life. Refusing to fill in a census return, declining to undergo electroconvulsive therapy, or greeting the rhetorics of politicians with scepticism may hardly be revolutionary activities, but they can represent resistance to the state none the less, and they give the lie to any assumption that the process of state formation is ever entirely uncontested.

The spread of modern states: statehood as aspiration

The apparatus of the modern state, complete with its complex geographies, differentiated institutions, and high levels of infrastructural power, was therefore emphatically not the product of a neat process of political development to ever-more progressive or 'democratic' forms. On the contrary, it represents the results of centuries of sporadic, *ad hoc* and unintentional developments. For much of the time war and the resourcing of war has been a key influence. More recently, the growth of administrative power, both through surveillance and population monitoring on the one hand and through the emergence of strategies for internal pacification and social control on the other, has been important. These processes have all involved particular uses of space and production of spaces, and all of them have to be understood as cultural transformations as well as political, economic and military ones. Modern, surveillant states are vast and powerful organizations which embody and act for particular social interests (including those of the dominant groups within the state apparatus). In the next chapter I consider in more detail what the modern state does and in precisely whose interests it operates. Beforehand, however, I want to close the present discussion with reference to the global impact of the *idea* of the modern state.

Imperialism is the subject of Chapter 4, but I will refer to it briefly here because of its impact on the process of state formation. There are two important ways in which imperialism influenced the development of states. First, during the rise of Europe's overseas empires, the imperatives and cultures of imperialism were of great importance in conditioning the formation of states *in Europe*. Second, during the period of decolonization, the territories and (European style) administrative apparatuses bequeathed to the newly independent areas by the departing imperialists were central to the formation of states *in the South*.

Colonialism was a crucial influence on the development of states throughout Asia, Africa and Latin America. Firstly, imperial administrations operated in territories which had been mapped out through the processes of colonization. The boundaries were partly the product of conflict between imperialist powers over territory, partly the results of conflicts between colonizers and the colonized and partly a consequence of the imperialists' desire to organize space so as to facilitate the exploitation of resources.

With the eventual decolonization (at least in a formal, political sense) of these territories, the newly independent states stepped into the administrative map of the colonizers, although it rarely bore any relationship to the social or political geography of the pre-colonial societies. This crucially weakened the capacities of post-colonial states in key areas of their activities. They also inherited a state apparatus which was culturally alien and which had been designed for the twin purposes of subduing the local population and facilitating the transfer of resources to the metropolitan core.[41] While this legacy hardly provided a propitious start for many newly independent states, there was in practice little that they could do other than adopt the model of the modern state, at least in broad terms.[42] One of the difficulties with this, is that (as we have seen) the modern state depends upon the provision of large resources in order to carry out its activities. The position of many post-colonial countries in the world economic system has ensured that at best such resources are limited. The elaborate edifice of the modern state is expensive, and supporting it has in some cases added to, rather than solved, the economic difficulties of poor countries. According to Ron Johnston the failure of states in the South to secure 'development' has generated a cycle of political instability in many areas,[43] which has typically led to more liberal regimes being succeeded by more authoritarian ones and vice versa.

However, it would be a mistake to overstate the extent to which European notions of the state have been imported into other contexts. The political scientist Jean-François Bayart counsels against seeing states in sub-

Saharan Africa as unstable, weak, ineffective and corrupt.[44] Such images, he suggests, are not only offensive, but also inaccurate. They are not failed versions of European states, fatally undermined by a combination of indigenous inadequacy and the global economic order. Rather, African politics and state formation must be understood in their own terms and be seen as ordinary and human, not as pathological deviations from some Western norm or ideal.

Having said that, it is unlikely that any society could avoid the general model of the modern state – bureaucratic, territorial, complex and militarized – since the pressures which generated it in Europe have to some extent become global in their scope.[45] The importance of Bayart's work is to point out that within this general model, states in different societies can take very different forms, and that those forms have to be understood as the products of their own histories and trajectories.

A further influential role for the *idea* of the modern state stems from a second great movement of the twentieth century: nationalism. The anti-imperialist struggles which led to the creation of independent states in the ways I have described were one form of nationalism and led in some cases to what has been called 'flag nationalism' or 'state nationalism': the attempt to develop a sense of nationhood and national belonging on the basis of nothing more than residence in the same state's territory.[46] Other nationalisms, grounded in various constructions of ethnic identity pursue the ideal of the modern state understood as a nation state: an organic synthesis of state and people reminiscent of Ratzel. In such campaigns a discourse of statehood is developed in which the 'destiny' of the 'nation' is presented as dependent upon achieving statehood – a territorial space, in which the 'community' of the nation can govern itself. Such arrangements are mythical of course – as we have seen, actual processes of state formation are not quite like that. But they are also extremely powerful as the nationalist war in the former Yugoslavia, which began in 1991, attests. These issues are pursued in more detail in Chapter 6.

I have concentrated in this chapter on the rise of the modern state in Europe. This has been at the neglect of issues of state formation elsewhere. I make only a partial apology for this. It is certainly the case, as Bayart's work makes clear, that state formation must be understood in its particular historical and geographical contexts. This is true, not just for Africa, but everywhere, including the crucially important states of China, Japan, southeast Asia, and Russia. All of these states have had distinctive processes of formation and must be understood as such. However, all states in the contemporary world are examples of the modern state form. They are certainly all distinctive, but the patchwork of territorial states which covers the whole land surface of the globe is a state system in which, through the mutual recognition of claims to sovereignty and a perpetual interchange of ideas and information about what states are for and what they do, the modern state has been universalized (at least discursively) as natural. Locating the origins of that form in the particular circumstances of a limited period and place helps to undermine its claims to naturalism, and thus in some small way questions its claims to authority too.

According to Charles Tilly, the modern form of the state has reached this state of ubiquity at precisely the moment that it has started to run into serious challenges to its claims to authority. In the next chapter we will examine in more detail what it is that modern states do in the contemporary world and whether the decline of the state has really set in.

Notes to Chapter 2

1 Friedrich Ratzel, 'The Laws of the Spatial Growth of States', in Roger Kasperson and Julian Minghi, eds, *The Structure of Political Geography* (London, London University Press, 1969), pp. 17–28, emphasis added. (Originally published in German as 'Die Gesetze des räumlichen Wachstums der Staaten', *Petermanns Mitteilungen* 42 (1896), pp. 97–107.)
2 Martin Glassner, *Political Geography* (New York, John Wiley and Sons, 1993).
3 Edward Soja, 'Communications and Territorial Integration in East Africa', *East Lakes Geographer* 4 (1968), pp. 39–57.
4 Ron Johnston, *Geography and the State: An Essay in Political Geography* (Basingstoke, Macmillan, 1982). Gordon Clark and Michael Dear, *State Apparatus: Structures and Language of Legitimacy* (Boston, MA, Allen and Unwin, 1984).
5 I use the term 'South' as a shorthand term to refer to (most of) the countries of Asia, Africa and Latin America. The countries and people of the South probably have more differences between them than they have things in common. However, they are similar in having been made objects of international policy and concern, which identifies them as 'poor', 'underdeveloped' or 'Third World'. Many of the terms used to discuss them (such as 'developing countries') derive from discourses constructed largely in the 'North' and by 'Northern' agencies, governments and academics. Rather than reproduce those discourses and their attendant power relations here, it seems preferable to choose a less tendentious term. Whenever the term 'South' is used in this way, however, it should not be taken to imply that its constituent countries are a homogeneous group.
6 Anthony Giddens, *The Nation-state and Violence* (Cambridge, Polity Press, 1985), p. 13.
7 For a discussion of the significance of 'claiming authority', see Chapter 1.
8 The main exception is Antarctica. While the governance of Antarctica raises interesting questions about states and territory, I will not be considering it here.
9 The same does not apply to most of the oceans, the sea bed, or the space above the earth's atmosphere.
10 Charles Tilly, *Coercion, Capital and European States: AD 990–1990* (Oxford, Blackwell, 1990), p. 2.
11 Giddens, *Nation-state*, pp. 49–51.
12 Nicos Poulantzas, *State, Power, Socialism* (London, New Left Books, 1978).
13 It is no coincidence that the words 'state' and 'statistics' share the same root.

14 Anthony Giddens, *The Constitution of Society* (Cambridge, Polity Press, 1984), pp. 8–12.

15 *Op. cit.*, pp. 6–7.

16 Michel Foucault, 'Governmentality', *Ideology and Consciousness* 6 (1979), pp. 5–21.

17 *Op. cit.*, p. 8.

18 The masculinist overtones of the metaphor of 'penetration' are clear. The gendering of the state is considered in the next chapter.

19 Note though, that preparing to fight wars continues as a routine activity.

20 Joseph Camilleri and Jim Falk, *The End of Sovereignty?* (Aldershot, Edward Elgar, 1992), p. 11.

21 Giddens, *Nation-state*, pp. 94–5.

22 Giddens, *Nation-state*, pp. 31–4.

23 Tilly, *Coercion*, pp. 14–16

24 *Op. cit.*, pp. 70–1.

25 *Op. cit.*, p. 87.

26 Christopher Dandekar, *Surveillance, Power and Modernity* (Cambridge, Polity Press, 1990), p. 57.

27 Tilly, *Coercion*, p. 64.

28 *Op. cit.*, pp. 39–40.

29 Giddens, *Nation-state*, p. 13.

30 Tilly, *Coercion*, p. 30.

31 *Op. cit.*, p. 30.

32 *Op. cit.*, p. 30.

33 *Op. cit.*, p. 15. Note that Tilly distinguishes between the national states (what I have called 'modern states') and nation states. The notion of the 'nation state' represents an aspiration for nationalist movements – a situation in which the boundaries of the state coincide with those of a culturally-defined nation. Few, if any, actual states fit this description. However, Anthony Giddens uses the term 'nation-state' to refer to modern states, since modern states usually *purport* to be nation states (for political reasons) and are widely regarded as such.

34 Philip Corrigan and Derek Sayer, *The Great Arch: English State Formation as Cultural Revolution* (Oxford, Blackwell, 1985).

35 Thanks to Miles Ogborn for this point.

36 Michael Mann, 'The Autonomous Power of the State: Its Origins, Mechanisms and Results', in *States, War and Capitalism* (Oxford, Blackwell, 1988), p. 5. The essay was first published under the same title in the *Archives Européennes de Sociologie* XXV (1984), pp. 185–213.

37 *Op. cit.*, p. 5.

38 Giddens, *Nation-state*, pp. 179–80. Giddens goes on to make the important point that social statistics are closely connected to the practices of social research, and the emergence of both from the eighteenth and, especially, nineteenth centuries led to the development of 'social sciences' which were not only 'about' society, but also implicated in the development of society: another example of the ways in which discourses produce their objects (see Chapter 1).

39 Giddens, *Nation-state*, p. 182.

40 On the significance of geography to these developments, see: Felix Driver, *Power and Pauperism: The Workhouse System 1834–1884* (Cambridge, Cambridge University Press, 1993); Chris Philo, ' "Fit Localities for an Asylum": The Historical Geography of the Nineteenth-century "Mad-business" in England as Viewed through the Pages of the *Asylum Journal*', *Journal of Historical Geography* 13 (1987), pp. 398–415.

41 Stuart Corbridge, 'Colonialism, Post-colonialism and the Political Geography of the Third World', in Peter Taylor, ed., *Political Geography of the Twentieth*

Century (London, Belhaven Press, 1993), p. 176.

42 Tilly, *Coercion*, p. 192.

43 Ron Johnston, 'The Rise and Decline of the Corporate Welfare State: A Comparative Analysis in Global Context', in Peter Taylor, ed., *Political Geography of the Twentieth Century* (London, Belhaven Press, 1993), p. 162.

44 Jean-François Bayart, *The State in Africa: The Politics of the Belly* (London, Longman, 1993). First published in French in 1989.

45 Tilly, *Coercion*, pp. 192–225.

46 Flag nationalism is not limited to the post-colonial world. UK nationalism is a flag nationalism. See, for example, Tom Nairn, *The Enchanted Glass: Britain and its Monarchy* (London, Radius, 1988).

|3|

Liberal democracy and the welfare state

Overview

In the last chapter I considered some of the processes through which the global system of modern states developed. I now want to examine the internal politics of states in a little more detail. To do this I will focus on one particular type of state: the liberal democratic welfare state. This form of the state is by no means universal, but it is important because it has often been held up as a model of political arrangements which provides both social justice and political freedom. As we shall see, such claims can sometimes ignore the inequalities and limitations of liberal democracy and the welfare state. In recent years, there has also been considerable academic and public discussion about the future of the welfare state, with some writers arguing that welfare states are prone to crises and may even be in a state of decline.

To begin with I will briefly locate the liberal democratic welfare state within the variety of state forms which exist in the modern world by summarizing the characteristics of other types of states. The remainder of the chapter then focuses in more detail on liberal democracy and the welfare state. First, I will outline an approach to understanding this form of the state drawn from the work of the political theorist, Bob Jessop. Jessop argues that it is not possible to develop an abstract theory of the state which can account for all its possible forms and features in detail. Instead, he proposes a 'weak' theory of the state which can guide concrete investigation. Jessop's weak theory is broadly compatible with the framework I outlined in Chapter 1. Second, I will consider the geography of liberal democracy by thinking about the relations between states, their citizens, and space. It is often argued that full citizenship can be exercised only if it includes social citizenship, that is, the right to a guaranteed minimum level of education, income and health care. The geography of welfare states is the third area I will discuss. Finally, I will turn to the future prospects for the liberal democratic welfare state and look at the challenges which are being posed to it both from within and from without.

Colouring in the map: the diversity of states

Types of states

While the modern state was forged in Europe through warfare, and military expenditure still accounts for significant parts of the budgets of most states, their bellicose origins and contemporary militarism are partly obscured, and to a degree mitigated, by the major growth of welfare expenditure and economic policy – those activities which were identified in the previous chapter as 'low politics'. The term 'welfare state' is used to refer to states which provide a variety of welfare services to their populations. Most often these include at least education, health care and a minimum income during old age, incapacity or unemployment. However, the growth of welfare states has been far from even. While some societies have elaborate and well-resourced welfare programmes, others have only the most token provision, or, in some policy areas, none at all. At the same time, the provision of welfare has been combined in different ways with policies aimed at generating economic growth, strategies for maintaining internal order and different kinds of formal political relationships between the state and the population.

Any attempt to produce a concise typology of states in the contemporary world is bound to be extremely crude, and to conceal as much as it reveals. All typologies produce groups of states which are internally heterogeneous and groups based on formal similarities are not much help for analysis. None the less, it is useful to undertake such an exercise even in a crude way, as it helps to reveal the orders of magnitudes (for example, of populations) involved in different political systems and state forms. The categories I am using are fairly conventional and descriptive and are intended merely to convey a sense of the diversity of state forms. They should not be understood to imply that the processes and causes of state formation are the same for all states in each group.

THE 'TRANSITIONAL' ECONOMIES

This group includes all but a very few of the former centrally-planned or state socialist countries, which are now making, or trying to make, a transition towards various forms of market-based economic systems. It must immediately be further divided into two sub-groups: those states which are also making a transition in their formal political systems as well (usually to a version of parliamentary representation), and those which have retained significant elements of the existing state apparatus, including the one-party system, which accords a leading role to a communist party or similar organization.

The first sub-group is largely constituted by the countries which made up the former Soviet Union and the states of Eastern Europe. In the case of the Soviet Union, a number of constituent republics (such as the Baltic states of Lithuania, Latvia and Estonia) are fully independent, while others remain in a loose confederation known as the Commonwealth of Independent States.

The second sub-group is now tiny in terms of numbers of states. However, it is a group which is not (yet) a complete irrelevance. Firstly, it was of enormous importance historically, and the historical legacy of various forms of Marxism–Leninism conditions the continuing development of the formerly state-socialist countries mentioned above. Secondly, it remains important because one of its members, China, is home to one-quarter of the population of the planet. During the 1980s there were significant shifts in the economic policies of the Chinese government, which introduced various market-orientated reforms. These were not, however, linked to any major political liberalization and the Chinese state remains highly authoritarian. Other countries in this group include the much smaller states of North Korea and Cuba.

THE WELFARE PROVIDERS

This group is also small in terms of numbers. It is made up of the so-called 'liberal democracies' of North America, Western Europe, Japan, Australia and New Zealand. The common characteristics of these countries are their material wealth, their highly complex capitalist economies, their relatively stable systems of representative government and their high levels (in absolute terms) of government expenditure on social provision including education, health care, social security and housing.

THE 'DEVELOPERS'

This category is a large and highly disparate group of countries, which are united only in the strategic orientation of their states to promoting social, economic and infrastructural 'development'. By 'strategic orientation', I do not mean that state policies actually produce 'development' in practice, but that it is their declared (and often overriding) goal to do so. We need to divide this group into four sub-groups.

1 *Newly-industrializing countries.* The so-called 'NICs' are united by the markedly rapid rates of increase in industrial production measured by changes in GDP per capita. Another small group, they are important because they provide a rather different model of development to the dominant Western one (although, like the Western model, it is unclear whether it could be replicated elsewhere), and because the state plays a particular role in promoting capitalist development through economic policy and internal pacification. They include the so-called 'Asian tigers' such as South Korea, Singapore and Taiwan.

2 *Oil-exporting countries.* Although this group includes some of the world's wealthiest economies in per capita terms, they are included here as 'developers' because of their reliance on the export of a single set of commodities: petroleum products. This has led to an emphasis in many cases on using the revenues from what is, after all, a non-renewable source of income to finance infrastructural and social provision, and, increasingly, economic diversification. The group includes Saudi Arabia and the other Gulf states.

3 *Early post-colonial.* This group includes those Latin American states that gained formal political independence from the European powers relatively early (by the mid-nineteenth century). Technically it also includes the United States. However, the Latin American group was marked by continuing economic dependence on Europe, and was not able to generate the same level of economic prosperity as its neighbour to the North. On the other hand, early political independence has allowed a much longer process of post-colonial state formation, and as a consequence state forms have had greater potential for autonomous change than the states in group 4.

4 *Later post-colonial.* This group is large in terms of numbers of countries and the population involved. It is also highly diverse, but its members are united by their relatively recent formal independence. They gained their independence in the wave of decolonization after the Second World War. Starting with India's and Pakistan's independence from Britain in 1947, the process is still not absolutely complete, although the remaining overseas possessions of the European powers are small in number and size. Much of Africa became independent during the 1950s and 1960s, although Portugal maintained colonial rule in some areas until the 1970s. The fact that colonial rule is so recent in these countries means that the process of independent state formation has had much less time to develop than in Latin America.

These categories are descriptive, rather than analytical. There is a degree of overlap, and the divisions between them are blurred. For example, China has an economy that is transitional between a centrally planned and a market system, but at the same time adopts a strategic orientation to economic growth, and is therefore also a 'developer'.

The apparatus and functions of states

I suggested in Chapter 1 that the capacity to pursue political strategies is dependent on resources. States have more resources than most organizations (although their resulting size and complexity makes them unwieldy when it comes to pursuing clear and unified strategies). In each case state resources are organized into a different 'state apparatus' – a set of organizations and institutions which undertake the functions of the state. What the state apparatus consists of varies from state to state, as do state functions. There are no functions or institutions which are inherently a part of the state, apart perhaps from the governmental legislature and core executive. Most things can be privatized, but prime ministers and presidents are probably exceptions! Also, in all modern states the cores of the standing armed forces and of the legal system are a part of the state (although some elements of them, such as defence lawyers, may be private).

Therefore the state apparatus does not have any fixed boundaries.[1] In some countries, health care is provided by the state, in others by the market and in many by both. Prisons are commonly state managed, but private prisons exist in the United States and in Britain. Schooling may be state or private, social workers may be employed by the state, or by private charities. Policing is usually a state function, but the growth of private security services, especially in North America, is blurring the boundaries here too. There are also major variations between the different groups of states in both the formal functions undertaken by states and the resources to provide them in practice. Many of the 'developers', for example, have states which formally undertake a wide range of tasks, but which are poorly resourced in terms of knowledge, organization, materials and staff, because the overall productive capacity of the economy is too low to support the full range of state functions in practice.

States and societies

There are also major differences between states in their formal political relations with their populations. In the 'welfare providers' group the dominant form of political organization is liberal democracy, a form of political representation involving the election of legislators and governments for a limited term through a universal franchise, usually with electoral competition between organized political parties. Other systems include:

1 various forms of dictatorship, military or otherwise, in which an individual or small group undertakes both legislative and executive functions without electoral accountability;

2 one-party systems, where the state apparatus is matched by a bureaucratic party organization and where there is an 'interweaving' of state
   and party functions and officials;
3 absolutist monarchies, which, while no longer a feature of European politics, continue in a number of countries, including a number in the oil-
   exporting group.

There is also considerable variation within these groups. The fact that
military regimes tend to be concentrated in poorer countries has led some
writers to suggest that this is a logical tendency produced in a political cycle.
This involves (a) the difficulties of generating economic 'development' leading to (b) a loss of legitimacy of the existing form of government and (c)
political instability which allows (d) a dictator to take power on the basis of
promises to restore order and thus to encourage investment. The repression
involved leads to (e) a loss of legitimacy for the new regime and, eventually,
(f) a return to democracy.[2]

Such arguments are helpful in identifying some of the different pressures
in different types of states. On the other hand, abstract tendencies are
worked out in very different ways in different contexts, and it is important
to recall Bayart's work on politics in Africa, which was discussed in Chapter
2, and which insists that states in Africa should not be seen as chronically
and inherently unstable as result of their colonial legacy. According to
Bayart, the widespread Western stereotypes of African states and societies
verge on racism by seeing them as institutionally weak, terminally corrupt,
and prone to tribalism and dictatorship. Instead analysts should understand
different ways of 'being political' as simply different, and not as pathological deviations from a Western norm.[3]

Institutions and strategies: a 'weak' theory of the state

'Strong' and 'weak' theories of the state

While there are significant variations between states, they all consist of a
mix of institutions, which are both the product of, and the medium for, the
development and pursuit of strategies and social practices. This relationship
between institutions and strategies was introduced in Chapter 1, but I want
to develop it in a bit more detail, and to strengthen it, specifically in relation
to the state. To do this I will draw on the 'strategic-theoretical' approach to
the state proposed by Bob Jessop.[4]

There is no universally accepted theory of 'the state'. According to Jessop, it is in any case impossible to develop a 'strong' theory of the state:

> A strong theory would provide an integrated account of the state in terms of a single set of causal mechanisms. It would explain all the institutional and operational features of a state in a given conjuncture. Even with the best will in the world, however, a strong theory could not be constructed. For it is simply impossible methodologically to develop a single, all-encompassing theory of so complex an entity as the nation-state in all its historical specificity.[5]

Instead, Jessop suggests, attention should be focused on developing a weak theory – a useful set of guidelines and principles which will assist analysis of particular states, but which do not assume that everything can be explained by a single set of mechanisms.

Modern states as 'institutional ensembles'

Jessop draws out the basis for a weak theory of the state from the work of two writers to whom I have already referred in previous chapters: Michel Foucault and Nicos Poulantzas. From Poulantzas he takes the idea that the state is an 'institutional ensemble' – the mix of institutional forms and practices which Poulantzas calls the institutional materiality of the state. For Jessop, the state may be defined as 'a specific institutional ensemble with multiple boundaries, no institutional fixity and no pre-given formal or substantive unity'.[6] This definition is echoed by some of the points that I have made in this and the previous chapters. It implies that there is no inherent division of functions between state and non-state institutions, that states are multifarious in their activities and that there are gaps, both geographical and organizational, in the extent to which society is 'penetrated' by the state, even in those functions for which the state is responsible.

The 'strategic-relational' approach

Jessop's strategic-relational approach to the state sees the state as 'the site, the generator and the product of strategies'. Writing from within a broad Marxist tradition of studies of the state, Jessop argues that earlier Marxist accounts fell into two broad groups.[7] First, there were those which emphasized the 'logic' of capital and which understood the state as an outcome of the laws and mechanisms of the process of capital accumulation. The second group stressed 'class struggle' and looked at the state as the product of the 'balance of class forces'.

Jessop argues that both approaches are unsatisfactory. The first approach often assumed that the state is automatically 'functional' for capital accumulation and inevitably the medium of 'bourgeois political domination'. Even when the state was seen as a problem for capital, this still assumed a single 'logic of capital' at any particular historical moment. Jessop insists that 'such assumptions are overly restrictive and ignore the scope for different accumulation strategies and the room for manoeuvre available to different class forces.'[8] On the other hand, the class-struggle approach, according to Jessop, tended to take the character of class forces for granted, and did not look at how different classes and class fractions arise and enter into relations of strategic alliance with, or opposition to, each other, or how very specific struggles affected the overall system:

> This presents us with a false dilemma. *Either* we emphasize the abstract logic of capital with its iron laws of motion, that is, its structurally inscribed tendencies and counter-tendencies. *Or* we concentrate on the overly concrete modalities of class struggle and cannot explain how such struggles tend to reproduce capitalism rather than produce a collapse into barbarism or a transition to socialism. There is little attempt to develop rich and detailed links between these approaches and this failure intensifies the problems of Marxist state theory. Yet the notion of strategy seems ideally suited to mediate between them.[9]

Jessop develops his strategic-relational approach to the state along three axes. First, the state system is seen as a site of strategy; as the place, if you like, where political strategies happen. However, access to this site is easier for some than for others, and the form of the state apparatus is more suited to some strategies than others. Second, the state 'is also the site where strategies are elaborated'. In other words state officials and thus elements of the state apparatus act strategically in their own right. Jessop points out that this may mean that different parts of the state are operating according to different (and maybe even opposed) strategies, and that therefore the 'formal unity' of the state is not matched by a 'substantive unity'. Third, the state apparatus and the conventional ways of acting of state officials and institutions are themselves the products of past strategies. 'Past strategies' may mean the strategies of previous state institutions and officials, or those of other political actors, in the economy, the wider society or the international arena.

This approach fits very well with the interpretative framework which informs this book. There are, however, a couple of further elements which I would emphasize. First, I propose that the state should be seen as many sites, rather than just one. This is entirely compatible with Jessop's stress on the lack of substantive unity of the state. Seeing the state as a multiplicity of sites raises issues about the spatial organization of the state apparatus.

Second, most (though not all) of Jessop's references to political strategies and social struggles are to *class* strategies undertaken by actors with particular class identities. However, there is no reason why the general approach should not include political strategies undertaken on the basis of political identities constituted in relation to gender, ethnicity, sexuality, locality and so on. Indeed, since contemporary social and cultural theory stresses that these identities and their attendant strategies cannot be reduced to, or subsumed within, class categories, there is every need to ensure that studies of the state are sensitive to the specificity of multiple identities and strategies.

Finally, Jessop's ideas imply that strategies can become 'path-dependent'. That is, while there may be a multitude of possible strategies open to an actor during a particular struggle, crisis or episode of state formation, once one particular strategy begins to be followed, the other options are made progressively less easy to adopt. This occurs because the chosen path involves specific forms of behaviour, frameworks of understanding and institutional changes which embed the participants into a particular course. Other strategies can be undertaken, but they will, as it were, run against the grain of state formation. Although I have borrowed the notion of 'path-dependence' from the literature on regional economic development, it is closely related to Jessop's notion of 'strategic selectivity'.[10] A good example of this is the directing of political strategies towards achieving change through parliamentary means (rather than revolutionary ones, for example). Once a particular system of government becomes established, it tends to condition future strategies as the strategic actors react to the context in which they are located. Where the parliamentary process is constructed discursively as the legitimate route for political change, and where the appropriate institutions exist, actors orientate themselves accordingly and adopt strategies which pursue reform through parliamentary means.

The reflexive monitoring of action and normative theories

Finally, the conditioning of strategies through existing sites brings up the issue of what Giddens calls the 'reflexive monitoring' of social systems. By this he means that, while strategies do not always have their expected outcomes, actors monitor the effects of their actions on a continuous (and often unconscious) basis. In the case of politics, this monitoring can become much more explicit. Indeed, as soon as the modern state began to emerge, political theorists began thinking and writing about what it was doing, and crucially, what it *should* do. This type of explicit monitoring by politicians and political philosophers is referred to as 'normative theory' because it proposes systems of 'norms' or rules about how things should be conducted. Thus, no sooner was the modern state born than it became the *object* of theorizing.

One of the principal concerns of early political theorists was that the modern state should not grow into a monster dominating its people, but should be limited and primarily concerned with securing the conditions under which people could be free to do as they pleased. These ideas are the core of liberalism, a normative theory which greatly influenced the early formation of the modern state, and which was the forerunner of the liberal democracies. I want to concentrate on the politics of the liberal democracies (which are now also the 'welfare providers' referred to above) in the rest of this chapter. This is not because the other groups are unimportant, quite the contrary. However, the liberal democratic states have been widely touted as normative models for politics elsewhere, and their activities have had profound impacts on the economic and political situation of most other states. In addition, in their twentieth-century incarnation as welfare states, they have faced increasing political and resource problems in maintaining the levels of social provision which their populations have come to expect. As a result, a growing number of commentators have started to argue that the welfare state is in crisis or reaching the end of its life, or even that the modern state itself is in decline.

States and citizens: the geography of liberal democratic states

Liberalism and the Enlightenment project

The discourse of the freedom of the individual is deeply ingrained in our dominant ideas about politics and societies in the West. Yet the individual view of human liberty is no more 'natural' than any other, nor is it a universally accepted ideal. In fact it was the product of a unique set of social, political and intellectual circumstances which came about in the eighteenth century and which we refer to as the Enlightenment.

THE ENLIGHTENMENT PROJECT

Along with urbanization, industrialization and the development of capitalism, the formation of the modern state has been a key element of 'modernity'.[11] As well as these shifts in social, political and economic geography, however, modernity has also involved a new set of human experiences and new ways of understanding the world and our places in it. Central to these new ways of understanding is a further process. It is sometimes referred to nowadays rather grandly as 'the project of the European Enlightenment'. In a narrow sense, the term 'Enlightenment' refers to the ideas of a group of eighteenth-century thinkers. More broadly, however, it has come to signify a wider intellectual movement which developed strongly after the Protestant Reformation. This wider movement valued rationality and human reason above superstition and unthinking religious observance and established a philosophical case for social change and social progress.

Understood narrowly, the Enlightenment refers to the work of the French *philosophes*, among them Diderot and Voltaire. Foremost of the works of the philosophes was the great *Encyclopédie* (Encyclopaedia) published in thirty-five volumes between 1751 and 1776. The writings of the *philosophes* challenged many of the traditional ideas of feudal society in Europe. Most importantly they were highly critical of superstition and conventional beliefs. They sought to establish human reason and rationality, and the possibility of scientific and social progress as the dominant world view (the Enlightenment is also known as the 'Age of Reason').

Understood more broadly, the Enlightenment refers to the wider development of scientific learning and rational thinking in the natural sciences, political economy and philosophy during the seventeenth and, especially, eighteenth centuries. The 'Enlightenment project', therefore, represents the various attempts to continue and to develop scientific reason and social progress in the period since, as well as the world view associated with them. Many writers now argue that in the late twentieth century this project has begun to falter. The experience of two world wars, the growth of environmental problems, the obvious lack of social progress in many parts of the world and the feeling that the world is becoming dominated by technology, rather than served by it, have all contributed to a questioning of the Enlightenment ideals. In addition a number of writers have argued that the Enlightenment faith in rationality gave special priority to one particular way of seeing the world: detached, all-encompassing and masculine. This, they argue, has led to a downgrading of other forms of understanding including those prevalent in non-Western cultures and those seen as feminine.

THE ENLIGHTENMENT AND THE INDIVIDUAL

The source of human reason is the human mind, and the Enlightenment project thus accorded human beings a central place in the world. Nowadays we rather take this for granted, but at the beginnings of the modern age it was much more common to place God at the centre of things, and to see human beings as but one part of divine creation (and a flawed part at that). Moreover, people were not treated as individuals, but as components of a social order, with the church, the monarchy and the nobility at the top; traders, merchants, teachers and doctors in the middle; and peasants and labourers at the bottom. The *philosophes* were hardly democrats themselves. However, their challenge to traditional ideas and forms of social organization led to their principles being adopted within the wider Enlightenment project by those who wanted to build human political freedoms on to the intellectual freedom which they had propagated.

The political thinking that grew out of the Enlightenment was quite new. In contrast with the ordered social world of medieval Europe, political philosophy began to emphasize individual liberty. The ancient powers of the absolute monarchs were decisively challenged, in some cases violently. In their place the beginnings of various forms of representative rule were established. These changes did not happen all at once, or to the same degree everywhere, but where they did occur they put into practice politically some of the philosophical ideas of the Enlightenment. These shifts are often labelled the 'bourgeois revolutions'. They challenged the old power structures based on inherited authority and the nobility or aristocracy. In their place the 'new class' of traders, industrialists and capitalists (the 'bourgeoisie') rose to power across much of Europe. Only the most radical of the political thinkers argued that *all* people should have a say in government. Most writers gave women no role in formal politics and in many cases working people without independent wealth were also excluded. However, although not *all* individuals were to be included, by the end of the eighteenth century individualism as an idea or a political principle had come to dominate thinking about government and economics, at least in northern Europe.

LIBERALISM

Individualism was the basis of liberalism. Today 'liberalism' is often used to refer either very narrowly to the ideas of a particular political party or very widely and rather vaguely to a tolerant or even permissive outlook. In the eighteenth and nineteenth centuries, however, its meaning was much more precise. In those days, 'politics' and 'economics' were not treated separately in the way they often are today. Instead, writers referred to themselves as 'political economists'. What we might call classical liberalism stressed the importance of individuals and individual freedom in both politics and economics, seeing the two as completely interlinked.

According to John Gray, a liberal political theorist writing today, liberalism is the political philosophy of modernity. He writes:

Liberalism – and most especially liberalism in its classical form – is the political theory of modernity. Its postulates are the most distinctive features of modern life – the autonomous individual with his [sic] concern for liberty and privacy, the growth of wealth and the steady stream of invention and innovation, the machinery of government which is at once indispensable to civil life and a standing threat to it – and its intellectual outlook is one that could have originated in its fullness only in the post-traditional society of Europe after the dissolution of medieval Christendom.[12]

This does not mean, of course, that liberalism has not been challenged. From different points of view, both conservatives and socialists have criticized the assumptions and practices of liberalism. However, conservatism and socialism are really responses to modernity (albeit ones which have fed back into it and conditioned its development). Liberalism, on the other hand, was central to the emergence of modernity in the first place.

Liberal writers included many of those involved in the French Enlightenment, as well as a group of Scottish writers who are often referred to as members of a Scottish Enlightenment. Among the latter were David Hume (1711–76) and Adam Smith (1723–90). They argued that the individual should be free from any interference from the government, unless it was necessary to protect others. This freedom included, most importantly, freedom to own property, to trade and to establish commercial enterprises. The assumption (and it is only an assumption) was that the best thing overall for society was to allow each individual to pursue their own selfish interests unhindered. This could best be achieved, the liberals argued, if the government was limited to the bare minimum of activities: just enough to allow individuals to carry out their own wishes unhindered by others (hence the close link between economic and political liberalism). This meant that classical liberalism did not guarantee, or imply, representative government. Liberal writers were concerned that rule by majority vote could result in restrictions on the individual freedoms which they valued above everything else.

Liberal democracy: the extension of formal politics

While there is no inevitable link between parliamentary democracy and the philosophy of liberalism, the pressures towards more representative forms of government came out of the same social and intellectual circumstances as liberalism. Many eighteenth-century political radicals, such as Thomas Paine (1737–1809) and the other American revolutionaries of 1776, and the French revolutionaries of 1789 based their calls for political freedoms on the same notion of individualism. In nineteenth-century Britain, liberals and political radicals called for the reform of Parliament to extend the franchise to a larger number of voters. This process began with the Great Reform Act of 1832, which extended the right of men to vote to a large part of the growing middle class. In 1848, popular protests demanding more representative government erupted across Europe, following further revolutionary upheavals in France which established the short-lived Second Republic and universal male suffrage. In Britain, the Chartists teetered on the brink of revolution before their demands for a radical extension of political participation finally failed.

In the following decades there were piecemeal expansions of the extent of formal political rights, although it was not until well into the twentieth century that the universal adult franchise was established throughout today's liberal democracies.[13] None the less, it was the movement towards more representative forms of government which mark the shift from liberal states to liberal democratic ones.

Charles Tilly argues that in the process of state formation rulers cede power selectively to other social groups (in the nineteenth century this mainly meant the bourgeoisie, and then the skilled working class) while bargaining for resources. The sentiment behind the American revolutionaries' slogan of 'no taxation without representation' was a powerful weapon in the struggle to secure full political citizenship for the widest number of people. The impact of pressures from outside the state apparatus for increased participation in the formal political process can thus be understood in relation to Bob Jessop's strategic-relational theory of the state. Social struggles for civil rights, and state attempts to resist them, bring different groups of state officials, ruling élites and popular protesters into strategic alliance and strategic conflict with one another. The outcome reflects both the differential resources available to the different parties, and the strategic trade-offs which groups are prepared to make in attempts to secure their own ends.

VOTERS, PARTIES AND ELECTIONS

In all liberal democracies, universal suffrage has been established on the basis of regular elections of parliaments, governments, presidents and so on with competition for electoral support being organized around systems of political parties. The precise form of electoral systems varies, but most of them are based on, or include, territorial units as electoral divisions. The existence of these geographical constituencies, and the convenient quantitative data generated by voting at elections, has generated a small, but lively, sub-branch of political geography concerned with analysing the geography of voting and electoral results.[14] Part of the interest of geographers has been in the spatial form of electoral districts, including the implications of the manipulation of spatial form for electoral advantage (gerrymandering) and the development of normative criteria on which to base proposals for effective, just or representative spatial units. While much of this research takes it for granted that liberal democracy is (in principle at least) the 'best' form of government, some geographers have developed more critical accounts which emphasize the role that the very existence of competitive elections plays in legitimizing the power of the state and its systematically unequal exercise of that power.[15]

The geography of voting is conventionally analysed using quantitative techniques. Such analyses work on the basis of accounting statistically for the distribution of one variable (in this case the choice of political party to support in an election) in terms of others (usually some combination of social factors, such as occupational class, sex, age and ethnic group). There are a number of technical difficulties with some of these approaches, which are dealt with more or less effectively in technical terms by quantitative analysts. In line with the agenda outlined in Chapter 1, however, my interest is in the potential contributions of social, cultural and political theory. To date these have received relatively little attention. There are at least two future avenues for investigation. First, statistical modelling of voting behaviour implicitly assumes that voting can be accounted for by a series of separate factors (such as class, sex etc.). However, an implication of much social theory is that human agency cannot be reduced to constituent factors in this way. Social relations such as class, gender and ethnicity are interrelated in complex ways through processes of mutual constitution, which cannot easily be captured in statistical models. The voting decision of a white, male, skilled manual worker cannot be reduced to a white ethnicity component, a male gender component and a skilled manual occupational component, because these various aspects of an individual's political identity and subjectivity are intimately bound up together. Furthermore, ethnicity and class are experienced in qualitatively different ways by men and women.

Second, one common finding of quantitative analyses of the geography of elections is that while much variation in voter behaviour can be accounted for statistically in terms of social attributes such as class, gender and so on, there frequently remains a systematic residual element related to location or region. Political geographers have frequently lighted on this pattern which, they suggest, shows that geography really matters. Typically, in a geographical area in which support for one party is stronger than for others, the dominant party's vote is higher than would be expected on the basis of the social make-up of the area alone. Geographers have taken this as an indicator of a so-called 'neighbourhood effect'. This assumes that voting behaviour is influenced by neighbours, so that individuals from any given social group will be more likely to support the locally popular party than members of the same group in areas where another party dominates.

However, insights of social and cultural theory suggest that the formation of political identities is a considerably more complex process than the political 'conversion by conversation' over the garden fence involved in the neighbourhood effect. This complexity is likely to increase as more and more formerly stable communities organized around traditional industries, occupations and ways of life are disrupted by dramatic economic and social change. Understanding the process of identity formation requires much more detailed attention to the nuances of discourse, ideology and symbolic practices, and a recognition that their geographies are not just local or neighbourhood-based, but are stretched across time and space through communications media of all sorts.

Although the subject has attracted relatively little attention, in comparison with elections, political parties also have a geography in at least two senses. First, as political institutions in their own right, individual parties are constituted over space in ways which do not simply reflect the geography of their electoral support, although this may influence it. Thus party membership, finance, organization and activity are all stronger in some areas than in others. Secondly, there is a geography to party systems. Thus in Britain the two major parties developed originally along lines of social class, whereas in the Netherlands the major party cleavage occurs along religious lines. These differences were discussed in an important paper by S. M. Lipset and Stein Rokkan.[16] Their ideas became known as the 'Lipset–Rokkan' theory, which related the observable geographical variation in party systems to the social structure of the country and the dominant social divisions arising from the processes of modernity. They identified two major processes associated with modernization and state-building: industrialization (involving conflicts between the agricultural and industrial interests and between employers and the working class) and state attempts at nation-building (involving conflicts across ethnic, religious and linguistic divisions). Depending on the combination of these various cleavages and their timing, one or more of them may come to dominate the party system. Because of the complex spatial variation in the relationship between different social conflicts and the process of European modernization, there are marked geographical differences between party systems.

Cultural politics and the spaces of citizenship

CITIZENSHIP RIGHTS: THREE DIMENSIONS

The right to vote in elections is often regarded as one of the defining rights of a citizen. Although the roots of the term and the concept lie in the city states of classical Greece and the empire and republic of classical Rome, the modern concept of citizenship developed in tandem with the development of the modern state. Its emergence was charted in a classic essay by T. H. Marshall in 1950.[17] Marshall argued that there are three aspects to the modern concept of citizenship, each of which involves a different form of rights: civil rights, political rights, and social rights. He suggested that, historically, the development of these different forms of rights matched the development of the English state from the liberal state of the eighteenth century, through the liberal democratic state of the nineteenth century to the social democratic welfare state of the twentieth century. Many contemporary writers would be more cautious than Marshall in assuming that there is a neat linear progression at work here, and would hesitate to apply the same evolutionary scheme to other states. Moreover, as we shall see, it would be a mistake to assume that the modern state involves the incorporation of its residents into full citizenship in any uniform way. However, while Marshall's account cannot be taken as a complete account of the relationship of actual states to their citizens, it is useful in clarifying the different dimensions that citizenship *can* involve.

The first dimension, *civil rights*, is closely linked to the liberal doctrine of the protection of the freedom of the individual. As we saw, liberalism argues that the state should be sufficiently limited in its powers so as not to restrict individual liberty unreasonably, but it should also be sufficiently effective to underwrite and protect that liberty from other threats. Thus the liberal state maintains military defences, institutions of law, order and justice, and protects the rights of its citizens to ownership of private property, to freedom of speech and to freedom from duress. The second dimension, *political rights*, involves the right to participate in the government of society, whether directly, through some form of participative democracy, or indirectly through the elections of representatives. The third dimension, *social rights*, involves the recognition by states that citizens have a right to a certain standard of economic and social well-being, which has involved the establishment of welfare and educational services of various kinds within a welfare state.

In accordance with the interpretative framework outlined in Chapter 1, it is important to see these aspects of citizenship as both the objects of social and political struggles and strategies, and the means by which those strategies are pursued. The extension of citizenship at each stage was not granted by the state without a fight, and involved demands from different social groups for the extension of rights. On the other hand, once the institutions of one form of citizenship were adopted, they could then be used as resources with which to campaign for other forms. Thus parliament (one of the sites of political citizenship) became part of the battleground in the struggle for social citizenship – welfare provision being introduced by Act of Parliament, for example. In addition, there is a 'discourse' of citizenship, which is used in, and produced by participants in, political struggles. Hence the state seeks to define discursively who is and who is not a citizen, and to insist that, for those who are included, citizenship is equal and universal. By contrast, those campaigning for civil, political and welfare rights use the same discourse of citizenship in arguing that certain groups are excluded from the benefits which citizenship brings. Both the meaning and the practices of citizenship are therefore changing and contested.

SPACES OF CITIZENSHIP

The fact that citizenship has no ultimately fixed meaning and established existence gives a clue to the important role of geography in constituting citizenship. At any one time there are limits to citizenship which are both formal and informal. These limits are the object of social and political conflicts over who is, and who is not, a citizen, and what kinds of rights they should enjoy. They are also spatial limits defined both by the borders of states and, in more complex ways, within states' territories. The stress on formal and informal limits is important, because although a wide range of social groups may be *formally* included in the legal or constitutional category of citizens, all too often there are *informal* mechanisms of exclusion and discrimination which mean that, in practice, certain groups are systematically unable to exercise their formal citizenship rights. This insight is the basis of the distinction between *de jure* citizenship (citizenship in law) and *de facto* citizenship (citizenship in practice). Within any political space, such as the territory of a nation state, *de jure* citizenship may well be enjoyed by a wide range of social groups, and, often, the majority of the population.[18] *De facto* citizenship, on the other hand, is often only fully available to a more limited range of groups. In some cases, *de facto* citizenship is exercised by people who are not *de jure* citizens.

Citizenship, both formally in law and informally in practice, is defined on the basis of inclusions and exclusions. The starkest of these is the right, enjoyed by *de jure* citizens, to residence within the territory of the state (inclusion) and the corresponding right of the state to exclude non-citizens from its territory, even in cases where those individuals have lived and worked as members of that society. The formation of secure state borders and practices of internal surveillance are central to the ability of the state to police *de jure* citizenship. For example, the political geographer Eleonore Kofman has argued that migration into wealthy European countries in recent years has been seen by their governments in terms of an invasion of people from the east and the south, and has led to an increasingly vigorous policing of the external borders of the European Union. Once inside the Union, non-citizens do not enjoy the same rights of free movement and social protection as citizens. In some European countries residents who arrived from elsewhere to fill labour shortages were, eventually, accorded *de jure* citizenship. In others, such as Germany, however, they were labelled as 'guest workers', with the state reserving to itself the right to exclude them, regardless of their length of residence or contribution to society.[19]

As well as the 'space of citizenship' marked out by the boundaries of the modern state, there are variations within states in the *de facto* citizenship enjoyed by its residents. In some cases groups who are not formally legal citizens and may not even be legal residents, are engaged in activities and organizations which call their exclusion from formal citizenship into question. Stephanie Pincetl has documented the strategies used by illegal Latino immigrants to Los Angeles County in California.[20] Although their presence is undocumented by the state, the Los Angeles economy depends upon their labour. In addition, their numbers are now so significant that their presence without political rights poses important questions for the American political system which was founded on the notion of participation and democracy. According to Pincetl, undocumented Latino residents have been extremely politically active, establishing political organizations and movements and gaining concessions from the state authorities despite their illegal status. This seems to suggest that in the particular geographical context of Los Angeles, a degree of *de facto* citizenship has been acquired by those who are technically non-citizens.

In other cases, groups of *de jure* citizens may find that the state limits the places and spaces in which they can enjoy *de facto* citizenship. The modern notion of citizenship appears to be based on principles of universalism and egalitarianism. In other words all members of the political community are assumed to enjoy the rights of citizenship and to do so equally. In practice though, as many feminist writers have observed, this model is based on some quite particular assumptions about the character of the archetypal citizen. The conventional model put forward by Marshall, for example, has been criticized for basing its notion of the citizen on masculine norms such as rationality, and the separation between a public political sphere and a private domestic sphere.[21] In virtually all the European liberal democracies, the right to vote was not extended to women until well after men, and in many cases married women had no political identity separate from their husbands. Recently writers on the politics of sexuality have extended this argument to gay and lesbian citizenship. Much of this debate has focused on the demands of gay men and lesbians for equal civil rights. For example, in Britain, gay political activists have campaigned for the equalization of the age of consent for heterosexual sex (currently 16) and homosexual sex (recently reduced from 21 to 18).

Arguably, one of the central tenets of liberalism, and thus of Marshall's civic rights, is the right to freedom from state restrictions in questions of private behaviour. In a recent legal case in Britain (arising from a police investigation known as Operation Spanner), a group of men were prosecuted (and convicted) for participating in sado-masochistic sex, despite the fact that those involved were consenting adults acting in private. According to David Bell, this case shows how the state can encroach on the citizenship of those whom it defines as sexually deviant, and that this encroachment has a key spatial component.[22] Before Operation Spanner, the spaces in which expressions of gay sexuality were tolerated by the state were already more confined that those of straight sexuality. In many areas of Britain two men holding hands or kissing in a public place may well find themselves subject to attention from the police. The state sanctioned what Bell calls 'dissident sexuality' only if it was expressed in private spaces. After the Operation Spanner case, however, dissident sexual behaviour even in private spaces is subject to state sanction. The geography of citizenship thus varies according to the ways in which the political identities of different social groups are constructed. The Spanner case illustrates how those who cannot, or who refuse to, conform to the culturally dominant political identity on which the modern notion of citizenship is based find that the spaces in which they are able to exercise *de facto* citizenship rights are often both highly constrained and highly unstable.

'From cradle to grave': the geography of welfare states

If political and civil rights were central to the emergence of citizenship during the eighteenth and nineteenth centuries, Marshall's final aspect of citizenship, social rights, developed later – from the end of the nineteenth century onwards. Social citizenship implies a relationship between states and individuals in which the former provides welfare benefits in cash or in kind for the latter on a formal basis and by legal right. These arrangements form what we understand today as 'welfare' states. In this section I will examine the development of the welfare state and its relationship to geography, while in the next I want to consider some of the processes which are undermining its dominance as the principal form of the state in wealthy industrialized countries.

Social discipline and social reform

The emergence of modern states went hand in hand with the development of large-scale urbanization and capitalist industrialization. States, dominant social groups, and prominent social reformers all participated in the production of two complementary discourses and sets of practices. The first of these was organized around the notion of social discipline. The rapid growth of cities and of a disenfranchised working class worried both the authorities and the newly wealthy urban bourgeoisie. They were concerned about the spread of disease and contagion in both a literal medical and a political sense. A discourse of social control and containment enabled, among other things, the establishment of social statistics as a distinct branch of government activity, the emergence of urban police forces and the development of a form of urban planning which gave priority to the opening up of the city to inspection by the state and, if necessary, control by the military.[23]

The second, related, discourse was one of social reform. The two discourses were related because, despite the clearly humanitarian intentions of at least some nineteenth-century reformers, the improvement of social conditions was widely seen as one way of avoiding the threats of disease and disorder posed by the rapid modernization of Western societies. In keeping with the liberal doctrines of the limited state, initial social welfare provision was limited to private and religious charity and the self-organization of working people into mutual aid societies, trade associations and labour unions. Once the franchise had been extended to the majority of the male population, pressure developed through the electoral system for increased provision of benefits and welfare services from the state.[24]

According to Christopher Pierson, the emergence of state welfare services dates from the last quarter of the nineteenth century. The introduction of key aspects of social insurance systems clusters in the last decade of the nineteenth century and the first quarter of the twentieth. In Europe by the time of the First World War, at least some of population in most countries were covered by some form of public social insurance for industrial injuries, health and old age pensions. In the United States, it was not until the economic crisis of the early 1930s and the New Deal promised by Roosevelt that similar schemes were introduced. However, the development of the welfare state was uneven not only in the timing of its introduction in different places, but also geographically in the *form* which it took in various countries and in its internal spread across space *within* countries. This allows us to identify three aspects to the 'spaces of welfare': different welfare regimes, geographical variations in the provision of services and benefits, and the emergence of local welfare states.

Spaces of welfare I: contrasting welfare regimes

In an influential book, Gøsta Esping-Andersen identifies three broad types of welfare state, which he calls 'welfare regimes'. He labels these the 'liberal', the 'conservative' and the 'social democratic'. These labels correspond to the dominant politics associated with the development of the welfare state in each case. As we have already discussed, the liberal state was founded on the basis of the freedom of the individual, with minimal state interference. While there is a sense in which liberalism is the principal form of politics in modernity, both conservatism and socialism were closely connected to liberal modernization as contrasting reactions. Conservatives tend to stress tradition, the importance of religion, the family, social status and social order. Socialists argue that the liberties of liberalism are illusory, because they are based on private property and the capitalist economic system, which tend to generate systematic inequalities and a social structure in which wealth is accumulated by the few at the expense of the many. Traditionally, socialists sought either to overthrow, or radically to reform capitalism. The use of (liberal) parliamentary means to achieve significant reforms, particularly in the welfare field, led to the so-called 'social democratic' state.

Esping-Andersen's account of the welfare state stresses three key components. The first is the degree to which welfare provision occurs outside the market, that is, in 'de-commodified' form. The second is the relative mix of market provision, state provision and family or community provision of welfare services. Following from these, the third component is that welfare states cannot be understood simply in terms of the rights of citizens to particular benefits, but must be analysed according to the ways in which the benefits are provided, and from what source. States do not divide rigidly into three groups but form clusters according to their mixture of liberal, social democratic and conservative features. Esping-Andersen outlines the liberal welfare state as follows:

> In one cluster we find the 'liberal' welfare state, in which means-tested assistance, modest universal transfers, or modest social-insurance plans predominate. Benefits cater mainly to a clientele of low-income, usually working class, state dependants. In this model, the progress of social reform has been severely circumscribed by traditional, liberal work-ethic norms. . . . Entitlement rules are therefore strict and often associated with stigma; benefits are typically modest. In turn the state encourages the market, either passively – by guaranteeing only a minimum – or actively – by subsidizing private welfare schemes. . . . The archetypal examples of this model are the United States and Canada.[25]

The second group is the 'conservative-corporatist' group which includes Austria, France, Germany and Italy:

> In these conservative and strongly 'corporatist' welfare states, the liberal obsession with market efficiency and commodification was never preeminent and, as such, the granting of social rights was hardly ever a seriously contested issue. What predominated was the preservation of status differentials; rights, therefore, were attached to class and status . . . the corporatist regimes are also typically shaped by the Church, and hence strongly committed to the preservation of traditional familyhood. Social insurance typically excludes non-working wives, and

family benefits encourage motherhood. Day care, and similar family services, are conspicuously underdeveloped; the principle of 'subsidiarity' serves to emphasize that the state will only interfere when the family's capacity to service its members is exhausted.[26]

Finally, the smallest group has the widest extension of the principles of universalism and de-commodification. In the social democratic welfare states,

> [r]ather than tolerate a dualism between state and market, between working class and middle class, the social democrats pursued a welfare state that would promote an equality of the highest standards, not an equality of minimal needs as was pursued elsewhere. . . . In contrast to the corporatist-subsidiarity model, the principle is not to wait until the family's capacity to aid is exhausted, but to preemptively socialize the costs of familyhood. The ideal is not to maximize dependence on the family, but capacities for individual independence. In this sense, the model is a peculiar fusion of liberalism and socialism. . . . Perhaps the most salient characteristic of the social democratic regime is its fusion of welfare and work. It is at once genuinely committed to a full-employment guarantee, and entirely dependent on its attainment. On the one side, the right to work has equal status to the right of income protection. On the other side, the enormous costs of maintaining a solidaristic, universalistic, and de-commodifying welfare system means that it must minimalize social problems and maximize revenue income. This is obviously best done with most people working, and the fewest possible living off social transfers.[27]

This case most closely approximates to the Scandinavian countries.

Esping-Andersen's arguments fit well with the approach to understanding state formation which I have outlined. For example, I have already stressed the importance of emphasizing the contrasting processes of state formation in different places. The relative balance of liberal, conservative and social democratic elements in different welfare states reflects the contrasting balance of social forces and the pattern of social struggles and political strategies adopted. Moreover, the development of welfare states is never completed but is an ongoing and shifting field of conflicts and alliances. Although Esping-Andersen does not deal with the impact of the cultural and discursive aspects of state formation in any detail, the significance of cultural processes in differentiating between the three regime types is implicit in his account. Thus the importance of the church and the family in the conservative cluster is perpetuated through a set of cultural understandings about the role of religious institutions, family life and the gender division of labour. In more liberal regimes, the role of the free market is not guaranteed, but must be continually reinforced against opposition through discursive strategies, while in social democratic systems, universalism, full employment and high levels of public expenditure are legitimated in part in terms of their material benefits, but also through discourses which construct them *as* benefits (rather than as costs, for example).

Spaces of welfare II: The geography of public services

We have already seen how the modern state is distinguished by its ability to reach across its territory in exercising its administrative power. One of the features of most welfare states is a political commitment by the state to use that capacity to promote a geographical universalism in the provision of public services. The declared intention is that, while particular welfare benefits and services may be available only to certain social groups (through eligibility criteria, such as means testing), access should not be influenced by place of residence. In other words, all those who are eligible for a specific welfare service according to national standards should be able to receive it, regardless of geography. In practice, however, things are rarely so straightforward. Although state provision is often more geographically even than market provision (which has inherent tendencies to uneven development[28]), the geography of public services is by no means wholly uniform.[29] In a simple sense, of course, the unevenness of public service provision reflects the unevenness of population distribution. What is of more significance, however, is the way in which the location of public services can produce more or less equitable outcomes in terms of social justice.[30]

A concern with territorial justice in the material outcome of public service provision is clearly of considerable importance, not least to the users of public services. In addition, of particular interest within the framework of this book are the ways in which a doctrine or discourse of territorial justice may be mobilized by different political actors inside and outside the state. Although it was not always couched in explicitly spatial terms, most welfare states have in the past stressed universal accessibility to public goods such as social services, health care and education. In many countries the planning of public services formalized political demands that public hospitals, schools, housing finance and so on should be distributed geographically in order to achieve wide population coverage. Recently, however, this discourse of geographical equity has been contested by the neo-liberal critique of the welfare state (see below). According to this perspective, the free market is capable, if unhindered by the state, of producing the greatest overall social benefit, while the welfare state has grown to the extent that it has crowded out the private sector, which is regarded by neo-liberals as the only genuinely productive part of the economy.[31] During the 1980s, many governments used parts of the neo-liberal argument in developing political strategies to 'roll-back' the welfare state, and in some cases this involved challenges to the concept of spatial evenness in public service provision.

In the United States, for example, the Reagan administration of the 1980s introduced sharp reductions in welfare finance provided by the Federal Government. Because of the characteristics of the federal system, in which considerable power and institutional capacity resides at the (lower) state level, federal programmes in a range of fields play an important role in underwriting a degree of geographical evenness, and thus spatial equity. If federal taxes and spending are cut back, then there is a net transfer of resources from poorer groups to richer groups in the population, and, by extension, from poorer areas to richer areas. At the same time, the remaining provision, at the state level, tends to generate further unevenness, because the political strategies of state governments and their resulting policies can vary quite widely.

The discursive elements of such changes are crucial, but they are not straightforward. It is after all an unusual government which proclaims that its intention is to increase social inequality! Rather, tax- and welfare-cutting programmes are organized through other discourses, such as 'the encouragement of investment', 'enhancing individual responsibility' or 'increasing flexibility'. In many cases such discourses contain an explicit geographical component. 'Flexibility', for example, may be couched in terms of 'giving increased flexibility' to elected local governments and local administrators, which implicitly sanctions greater geographical variation in provision.

Neo-liberal thinking is often linked with rational choice theory, which is applied to the provision of public goods and services as 'public choice theory'.[32] According to this perspective, individuals act rationally and in their own self-interest to maximize their welfare in terms of the mixture of public services they receive and taxation costs they bear. One geographical application of this was proposed by C. M. Tiebout. Tiebout argued that there should be a large number of small local government units each providing a different mixture of taxes and services.[33] If the assumptions of rational choice theory hold, then individuals would move to the local government area which provided the particular mix of taxes and services which maximized their welfare. Such a scenario clearly depends on the decentralization of political decision-making about taxation and public expenditure to the local level and actively promotes unevenness in the provision of public services. In Britain in the 1980s the Conservative government of Margaret Thatcher introduced a new form of local taxation (the community charge, which became widely known as the 'poll tax'). Initially this was promoted through a discourse of local accountability and consumer choice which owed much to the thinking behind the Tiebout model.[34]

Spaces of welfare III: the local state

Geographical unevenness in the provision of public goods can thus arise either because the service involved can only be provided in a locationally specific way or because it is distributed unevenly as a result (whether intended or not) of public policy. The third of my spaces of the welfare state overlaps with the second, since, as we have seen, many local services are provided by decentralized units of local government. For states with relatively large territories (in which most people live), the logistical and administrative problems in delivering public services to a geographically scattered population leads to the formation of more or less localized systems of administration and service delivery. In a great many cases at least part of the local apparatus of the state includes local elections and decision-making about the provision and co-ordination of certain state activities at the local level by local politicians.

The term 'local state' was coined by Cynthia Cockburn in 1977 in her analysis of the politics of the consumption of local government services in the London Borough of Lambeth.[35] In principle, however, the term can be extended to apply to all sub-national parts of the state apparatus, even where these are wholly dependent on the central state authorities (for example locally-based agencies implementing central government policies, such as the British Urban Development Corporations). Systems of local government and administration vary quite markedly between different countries. In federal systems, such as those in Germany and the United States, the central federal state is made up of a number of smaller units (*Länder* in Germany, States in the US) which have considerable constitutional autonomy from one another and from the federal government. The components of a federal system usually have tax-raising and legislative powers and their own parliamentary or other representative governments. They also usually have their own internal administrative geography, which may also involve elected authorities. These lower tiers are normally subordinate to the governments at the State (*Land*) level, rather than to the federal government. In unitary systems, such as that in Britain, by contrast, the autonomy of the sub-national state institutions is much more constrained. In so far as they have decentralized power, these are devolved to the local tier by the central state, and could, in principle, be revoked. Such a recentralization has, in fact, occurred in respect of a variety of issues in Britain during the 1980s and early 1990s.[36]

According to Gerry Stoker, Britain has also seen a considerable growth in the role and extent of what he labels 'non-elected local government', sometimes at the expense of elected local agencies.[37] However, wherever the precise line between the elected and non-elected local state lies, the existence and role of elected local agencies remains politically important. Although in large territorial states, administrative decentralization will be required for convenience, there seems no inherent necessity for this to be elected. According to Gordon Clark and Michael Dear, the widespread existence of an elected element to the local state acts to buttress the legitimacy of the state as a whole by allowing the state at all levels to claim local account-ability.[38] They also stress the significance of a discourse of self-determination in local electoral politics.

Like all social institutions, the apparatus of the local state is the site of production of competing internal discourses. Since state institutions are not entirely homogeneous organizations, there are continual *internal* political conflicts. These clearly vary, depending on circumstance. In Britain in recent years, many internal conflicts have focused on new forms of management practice. The so-called 'new public sector management' has drawn heavily on certain versions of private sector management thinking including those which stress quality and customer care. According to the same management discourses, attempts to introduce changes into organizations should involve cultural as well as organizational restructuring. This management thinking tends to assume that there is a single institutional culture, in which all employees or members of the organization share the same institutional values and outlook. While the interpretative framework I have been using shares this emphasis on cultural understandings, it also stresses that it is highly unlikely that there will be a unitary or homogeneous set of cultural norms in any large institution, and moreover, cultural meanings and values are just as likely to be the focus of conflict as any other aspect of an organization.[39]

Challenging the liberal democratic welfare state

Having outlined some of the key ways in which the welfare state is geo-graphical in more than merely its territorial extent, I want to turn to some of its contemporary problems. These are of two (closely connected) sorts. First, there are problems which arise, according to some writers, from inher-ent tendencies and logical contradictions within the apparatus and func-tioning of the welfare state and in the relationship between state and society. Second there are challenges to the welfare state from particular political and social groups, particularly those who feel excluded from the full rights of social citizenship on which the welfare state is based.

Challenges to the state come from a whole variety of different sources and from all points on the political spectrum. The sheer size and power of the state apparatus, and the multifarious character of state functions and roles, makes it almost inconceivable that it could act always in ways which please everyone. When we add to that logistical constraint, the surveillance and monitoring functions of the state, the fact that state resources are dis-tributed in systematically unequal ways, and the cultural formation of the state in ways which exclude certain groups, the potential for opposition to the state is large. This does not mean that welfare states have not achieved increases in welfare for large numbers of their citizens. On the other hand, it does mean that we need to take seriously Poulantzas' injunction that the state is the site and product of social and political struggles, and not some kind of neutral observer, above the fray.

The contradictions and crises of the welfare state

Because I have wanted particularly to stress the administrative, social and cultural aspects of state formation, I have not so far paid much attention to those approaches which focus on the political economy of the welfare state. It is important, because it details the relationship between the state and the capitalist system of production. Although a huge range of state activities contribute either directly or indirectly to the reproduction of capitalist social relations and to the process of accumulation of capital, the state itself rarely engages directly in production for profit. Indeed, bearing in mind the liberal heritage of the modern state, one of its claims to legitimacy rests on state intervention being used only in supposed cases of market failure, and not becoming a general principle across the broad range of economic production. This means that modern welfare states are almost wholly dependent for resources on production in the private sector. Even where finance is borrowed by the state, for example by issuing government bonds, the commitment to repayment implies, at least in principle, an ability to obtain future resources from the private sector.

Financial resources may be of two types: either a direct tax on private profits, or an indirect one, for example through the taxation of wages or popular consumption. Capitalist production thus pays for the state, and also gains from it, as the state provides infrastructure, a legal system, an educated and healthy workforce and so on. However, the size and complexity of the state apparatus, and the political strategies of dominant political and bureaucratic groups within it, means that there is no direct and straightforward functional relationship between the state and capital accumulation.

As we saw in the previous chapter, the modern state did not develop for the purpose of supporting capitalism; rather, the development of capitalism in particular places allowed certain states to acquire the resources with which to undertake other activities (such as waging war). The process of political bargaining for access to resources certainly involved the state increasingly in policies which furthered the interests of at least some sections of capitalism, but the state does not exist because capitalism needs it. Since the state is an arena of social struggle between a wide variety of social interests, other groups besides those of capital can and do make demands of, and gains from, the state. The establishment of the British National Health Service, for example, was bitterly opposed by the private medical establishment and other sections of capital. While some of its activities feed into the capitalist system by providing healthier workers, it is difficult to see how others, such as the care of elderly people, or of those suffering mental distress, has any significant benefit for the process of capital accumulation. The fact that the state does engage in such non-functional (for capitalism) activities shows the extent to which it is able to pursue its own agendas and political strategies.

Nevertheless, if state policies did have a consistently negative impact on the overall capacity of the capitalist sector, in the medium to long term there would be considerable negative consequences for the state itself. First, it will be faced with increased levels of social problems as a product of economic decline and the resultant increased costs. Second, it will find its own resource base stagnating, or even shrinking. The simple solution to this might appear to be for the state to attend extremely carefully to the requirements of capitalist production. However, such a strategy has its own difficulties. First, it will encounter opposing strategies within and outside the state apparatus. Second, there is no one singular capitalist interest which might be prioritized. In Britain, for example, many commentators have written on the conflicting interests of finance capital based in the City of London and manufacturing capital.[40] If the state is to prioritize the needs of capital, which section of capital should it target? Third, one of the functions that the welfare state has characteristically fulfilled has been the mitigation of the conflicts inherent within the capitalist mode of production, although in some cases this has been at the expense of perpetuating conflicts along other social divides, such as those of gender and race.[41] However, the problem is that such conflicts are endemic in modern societies and cannot be permanently resolved by the state, but only postponed.

The state is thus the focus of contradictory pressures. If it undertakes strategies to try to resolve difficulties in one sphere, it tends to exacerbate them in others. Some commentators have argued that the state is thus dogged by inherent crisis tendencies. Three of the most important writers here are James O'Connor, Jürgen Habermas and Claus Offe.[42] Although there are differences of emphasis between their accounts, their approaches are broadly compatible, and I will rely on Offe's account of the 'crisis of crisis management'. According to Offe the modern state in capitalist societies is marked by tendencies to *fiscal crises*, *rationality crises*, and *legitimation crises*.

FISCAL CRISES

A tendency to *fiscal crises* arises because the budget of the state has a tendency to grow more quickly than its resource base. For Offe, the political crises of the welfare state are displaced economic crises. As capitalist economies expand, he argues, they tend to become organized more and more by the state. This does not mean necessarily that the state takes over the actual management of firms, although this can happen. Rather, the continued reproduction of the capitalist relations of production increasingly depend on state activities in other areas, such as the regulation of the banking and credit systems, the organization of markets, the stimulating of demand, the provision of physical infrastructure and investment in human capital such as education and training. The paradox at work here is that as this process proceeds, the state has a tendency to absorb a greater and greater proportion of the gross national product, leaving a declining proportion for profit-making investment by the private sector and the process of capital accumulation:

> Budgetary decisions concerning revenues and expenditures have the double function of creating the conditions for maintaining the accumulation process as well as partially hampering this accumulation process by diverting value from the sphere of production and utilizing it 'unproductively' in the capitalist sense. ... This contradictory process can be seen as analogous to that of physiological addiction: the addict requires ever larger drug doses at the same time as the potential withdrawal phenomena that would follow a reduction of these doses become more and more crucial.[43]

RATIONALITY CRISES

According to Offe, 'administrative rationality ... is the ability or inability of the political-administrative system to achieve a stabilization of its internal "disjunctions"'. In order for the state to act rationally it must fulfil five criteria. First, it must maintain an operational distance from the immediate political demands. Second, it must be able to separate internally those functions which are concerned with economic management steering from those involved with securing legitimation or mass loyalty. Third, it must be able to co-ordinate its various agencies to prevent them acting in contradictory ways. Fourth, it must have, or have the means to acquire, sufficient information on which to take decisions. Fifth, it must be able to forecast adequately over the same timescale on which it aims to plan its activities.

The tendency to rationality crisis arises when the state is chronically unable to meet these criteria. Given my previous discussion about the porous and differentiated character of the state apparatus, it should come as no surprise to discover that in general, welfare states are rarely able to meet these criteria, and thus exhibit tendencies to rationality crisis.

LEGITIMATION CRISES

Finally, the *legitimation crisis* of the state develops because the state has a chronic inability to secure, in any stable way, mass loyalty for its activities. Offe identifies five problems here too. First, the welfare state is based on a commitment to provide high (and often rising) levels of social welfare for its population. Because this is an avowed aim of state policy, any failure to maintain welfare outputs has a greater impact on public support than failures in policy areas which are not a central and defining feature of the state. Second, the symbolic resources which might promote social integration increasingly decline as pre-industrial ways of life are eroded. Third, there are contradictions between social and cultural norms and understandings which destabilize the political culture. For example, the Protestant work ethic is in conflict with a growing hedonism. Fourth, as symbols promoting social cohesion and integration increasingly become drawn into the marketplace, they lose their validity as foci of popular solidarity. Finally, as the state draws more and more aspects of social life out of the market ('decommodification') people's working lives are less and less subject to the discipline of the labour market (for example) and their expectations consequently rise.

SUMMARY

To summarize, Offe argues that there are inherent tendencies to crisis in the welfare state because its activities undermine inputs which are essential to its survival (namely fiscal revenue and mass loyalty), and its internal decision-making processes are inadequate to the growing range of tasks which it attempts to undertake. Although Offe does not emphasize the role of practice and strategy explicitly, the tendencies towards crisis he identifies may be understood in my perspective as the unintended outcome of strategies pursued for other purposes. For example, institutional strategies intended to generate legitimacy, such as the provision of welfare benefits in line with popular expectations, may have the unintended consequence of contributing to the development of a fiscal crisis, through the resulting increased demand for taxation revenue.

It is important to stress that the processes identified by Offe are only crisis *tendencies*. An actual crisis (leading for example to the complete breakdown of the welfare state and its replacement by something new) will probably only develop if it is constructed politically as a crisis through argument and discourse. It is to these directly political challenges to the liberal democratic welfare state which we now turn.

Challenging the state: constructing critiques

Offe's arguments are compelling, but they are constructed at the level of inherent, logical tendencies in the structural relationships and typical dynamics of the state. They cannot explain why particular states at particular times come to be the focus of certain kinds of political strategies, or exhibit a particular concrete form of crisis. These issues can only be addressed by returning to a focus on the discursive constitution of the political and strategic political behaviour on the part of situated social groups.

Bob Jessop's strategic-relational approach (outlined above) to the state is useful here. In that approach, Jessop argues that the state is itself the site of conflicting strategies between and within social groups, political parties and state employees. All of these are themselves more or less institutionalized, and the success of different strategies has varied markedly. In Chapter 6, I will show how different social movements are able to achieve differing levels of political success according to the package of resources which they can command. The same applies here. Different strategies within and towards the state have different effects depending in part on the range of organizational, material and discursive resources which actors command. Here, though, I want to focus on the discursive side and look at the critiques which have been developed from different perspectives in relation to the liberal democratic welfare state. Such critiques are numerous and highly varied. They have been developed from a range of political perspectives. Here are the main elements of some of the more important arguments.[44]

NEO-CONSERVATIVE CRITICISMS: THE PERMISSIVE STATE

The social democratic welfare state, which places such store on the provision of welfare benefits and social rights, has been attacked strongly during the 1980s from the right of the political spectrum. In many countries after the Second World War, there developed a broad political consensus between right and left. Although the level of agreement should not be overstated, in many places there was a general acceptance of the existence of the welfare state. During the 1980s, this consensus broke down and welfare states became of the focus of significant critiques and restructuring, which originated among right-wing thinkers, but became more general, as even nominally socialist or labour governments (for example, in France and New Zealand) began to move towards more market-orientated provision.

However the right-wing critique was itself contradictory, combining both conservative and liberal arguments. So-called neo-conservatives attacked the welfare state for its supposed role in undermining the traditional structures of the family, religious life and voluntary self-help. In Britain and the US, for example, neo-conservative thinkers suggested that state welfare benefits systematically encouraged young, single women to have children without getting married, or establishing a secure home environment. According to this point of view, the state provision removed the need for a male breadwinner, and thus for the conventional nuclear family.

Such arguments have proved enormously controversial. Many opposing views have been put forward, suggesting among other things that the state is responding to wider social changes (in family structures, for example) rather than producing such changes, and that the conventional nuclear family need not be regarded as the only legitimate, or even the best, household form.

NEO-LIBERAL CRITICISMS: THE NANNY STATE

Despite their combination in the thinking of many right-wing political parties, liberal views are in many ways at odds with conservative ones. For example, a neo-liberal perspective would stress the freedom of people to live in any form of household which they might choose. While liberalism would agree with conservatism that the state should not make welfare payments to single mothers, it would extend the principle to all welfare benefits, arguing that the state provision of welfare should be cut to the bare minimum and that the state should not engage in *any* kind of 'social engineering', whether in favour of or against the traditional nuclear family.

In general neo-liberalism prioritizes market provision over all other forms and is critical of what it sees as the 'nanny state', protecting and cushioning households and companies from the rigours of market relations. The assumption of neo-liberalism is that the free market, left to its own devices, will secure the 'greatest happiness for the greatest number', while state intervention is bound in the long run to fail, or to provide welfare outcomes which are less efficient than those available through the market. In contrast to these arguments, critics on the left argue that the operation of the market does not tend to produce equilibrium conditions, or social justice, but operates in systematically unequal ways leading to greater inequality and, for those at the bottom, increasing hardship.

FEMINIST CRITICISMS: THE GENDERED STATE

Feminists have criticized a wide range of state activities on the basis of its gendered characteristics. There is considerable diversity within feminist writing. For example, it is common, although increasingly rather crude, to identify three main strands within feminist thought. These are liberal feminism, socialist feminism and radical feminism.[45] All varieties of feminism focus on the unequal position of women and the feminine relative to men and the masculine. Each strand is itself heterogeneous, but very roughly, liberal feminism focuses on the struggle for equal rights for women within the existing social system, while socialist and radical feminists argue that the existing social system is the source of gender inequalities and needs to be radically transformed. Socialist feminists suggest that capitalism plays an important (or even an overriding) role in producing gender inequality, while radical feminists stress the role of patriarchal social relations. This has led to socialist feminists focusing more on issues such as the position of women in the labour market and the role of women's domestic labour in supporting capitalist production, and to radical feminists emphasizing the role of men and masculine ideology in exploiting and oppressing women through (among other things) control over their bodies and through violence. There are numerous ways in which these three strands relate to different aspects of the activities of modern states. For illustration, I briefly outline three examples.

The role of male violence is important in the state in a number of ways. First, in the last chapter, I discussed how the formation of the modern state was intimately connected to war. Modern states are also highly militarized. It is clear that the overwhelming majority of the armed forces of modern states are men and that throughout history wars have been almost always fought between men. According to many radical feminists, preparing for and waging war are activities which are heavily gendered masculine, a gendering which is reinforced in modern Western culture through popular media, such as Hollywood war films. At the same time, it is often women who are left to pick up the pieces of war (through support on the home front, for example) and who are directly affected as civilian casualties. Press coverage of the war in the former Yugoslavia has stressed the extent to which territorial conquest is often followed by the sexual domination of women (through rape) by the victorious forces.

Feminists have also argued that the welfare activities of the state are gendered to the advantage of men. This varies from state to state, and Esping-Andersen's work emphasizes the ways in which welfare benefits which help women have been introduced to a greater extent in the social democratic regime cluster than in the conservative regime cluster (see above). In many cases, however, state welfare provision is supplementary to, and assumes the continuation of, women's domestic caring activities. In Britain, for example, the architect of the welfare state, William Beveridge, was clear about the gendered assumptions on which the British system was based:

In any measure of social policy ... the great majority of married women must be regarded as occupied on work which is vital though unpaid, without which their husbands could not do their paid work and without which the nation could not continue. In accord with the facts the plan for Social Security treats married women as a special insurance class of occupied persons and treats man and wife as a team. ... That attitude of the housewife to gainful employment outside the home is not, and should not be, the same as that of the single woman. She has other duties . . .. Taken as a whole the plan . . . puts a premium on marriage in place of penalising it.[46]

Furthermore, some commentators have stressed the ways in which bureaucratic forms of organization, such as those typical of welfare states, are gendered.[47]

Thirdly, liberal feminism, which stresses legal rights and equality of opportunity, has emphasized the state's role in providing (or failing to provide) the legislative and legal frameworks to ensure that women have the same formal access to jobs, political participation and the public sphere as men. Thus in many countries, governments have enacted equal rights legislation, making it illegal for prospective employers to discriminate between employees on the grounds of sex. A key focus here is the notion that citizenship, in its various forms, has not been universal in the past, inasmuch as it tended to exclude women, either deliberately, or by default.

Finally, it is important to stress that the gendered character of the state in all its forms has been one of the main foci of resistance from women. The growth of feminism has been one of the most important political, as well as intellectual, developments of the twentieth century. In the final chapter I will consider the geography of social movements, including the women's movement in more detail.

SOCIALIST CRITICISMS: THE BOURGEOIS STATE

Feminists have emphasized one axis of domination in which the state is implicated, that of gender. Socialist arguments also focus on domination, but tend to stress its class character. The socialist critique of the welfare state holds that the state operates in favour of the middle class and in the interests of capital to the detriment of the working class. Some socialist writers have stressed the ways in which the key decision-making roles in the state apparatus tend to be filled disproportionately by the bourgeoisie. Others have argued that the state is simultaneously the site, medium and outcome of class struggles.

Charles Tilly argues that the process of state formation is the outcome in part of the development of class alliances.[48] The same might also be said of the modern welfare state. While it is able to provide welfare services, such as health and education, to a broader cross-section of the working class than ever before, it also acts in the interests of capital, as I showed in the discussion of Offe's work, above. One of the ways in which the state acts for capital is through a discourse which seeks to identify state policies with the national interest. According to the socialist critique, the national interest is in fact the interests of capital in disguise.

States also base much of their claim to authority and legitimacy on economic policy. Given that the economic system is dominated by capitalist relations of production, which, according to socialist analysis, depends on the perpetual accumulation of capital and class exploitation, the promotion of economic growth is inevitably the promotion of the interests of capitalism. The picture is complicated, however, because in advanced capitalist economies, the class structure is much more complex than a simple division between the capitalist class and the working class. Much of the means of production is owned by large institutional investors, whose funds come ultimately from savings accumulated by ordinary employees (for example, in pension funds). Control of the means of production, however, is largely vested elsewhere among senior management and corporate executives. Many workers occupy intermediate positions in the spectrum of possible class relations, particular those who work in service functions, such as banking, insurance and advertising. Such employees are not straightforwardly members of a capitalist class, but they service the needs of capital. At the other end of the social scale, most advanced capitalist countries now have a large proportion of their populations in a state of permanent or semi-permanent un- or under-employment. Again, such people are not straightforwardly members of a unified proletariat, but at various times may be more or less excluded from the social relations of capitalist production altogether.

The links between the state and these various relationships are complicated. On the one hand the state's resources stem ultimately from the production undertaken by the capitalist sector. On the other hand, at any particular point in time, the state can pursue a variety of strategies in relation to different social groups. This 'relative autonomy' of the state provides scope for class-based (or indeed other) strategies of resistance.

Traditionally, class-based resistance to the state came through the formation of labour movements. With the growth of welfare provision came the growth of the state's own workforce. This meant that the state was increasingly involved directly in the constitution of class relations through its own role as an employer. Challenges to the state by the labour movement could (and in many cases did) now come not just through the political system (for example through the growth of labour and socialist parties) but also through industrial actions on the part of state workers.

THE CASE FOR THE DEFENCE: STATE FORMATION AND SOCIAL STRUGGLE

All of these perspectives, as well as critiques based on anti-racist and environmentalist thinking,[49] might suggest that the state has few friends left.

However, while the state is the object of critiques from a variety of political positions, the social and political movements based on those critiques seek, for the most part, changes in the state, rather than its abolition.[50] In other words, the construction of alternative discourses and political strategies usually represents an engagement with the state, rather than a complete rejection of it. As I have suggested, states are not static, finished products. Rather they are subject to continuing processes of formation and development. Moreover, states are the products of social struggles and alliances, which will shape their development and future formation. This suggests that while the state is changing, it is not necessarily disappearing. The future characters of particular states depend on the political success of different strategies and social movements of which these critiques are the basis. Politics is a fluid, unpredictable and contingent process, and it is impossible to foretell exactly where the liberal democratic state is going. Some tendencies are clear, however, and in the final section I will sketch some of the potential alternative futures for the state which may emerge.

The future for the state

CHALLENGING STATE SOVEREIGNTY

So far in this chapter, I have focused principally on the internal dynamics of states. In other words, I have examined the logical contradictions and immanent crisis tendencies in the apparatuses and functions of individual states and the critiques which have been developed by groups within their territories. There are also, however, important external challenges for states. Indeed these have become so marked in recent years, that many writers are talking of the decline of the state as a result.

As I have noted previously, states claim sovereignty. That is, they claim to be the highest legitimate political authority with respect to their territories. The critiques of the state which I have outlined seek to challenge the ways in which the exercise of state sovereignty favours particular interests at the expense of others. Often, however, they do not call into question the sovereignty of the state itself. Other changes do.

First, the globalization of economic processes undermines the ability of the state to plan, steer and govern the national economy. Increasingly, multinational corporations, international financial services companies and multilateral financial institutions such as the World Bank have been able to take decisions with little or no regard to the wishes of individual governments. The room for economic manoeuvre of governments and the scope for macro-economic intervention has been significantly diminished.

Second, we have seen the growth of new forms of political authority and governmentality which operate alongside state governments. Many international political institutions are wholly intergovernmental. That is, they operate, at least in theory, on the basis of mutual agreement between participating states. The United Nations and its various agencies are perhaps the most developed example. In practice, however, power within intergovernmental organizations is often not distributed evenly between participants (as a result of the differential distribution of resources). In the case of the United Nations, for example, the role of the United States has been central to many of the policies it has adopted, particularly where military intervention in an individual state is concerned.

Third, there are new forms of political authority which challenge the state's claims to sovereignty directly, because they are truly supra-national in character. In contrast with intergovernmental organizations, supranational bodies have taken over from states certain functions which used to be fulfilled by individual governments. The most significant example of a supra-national organization is the Commission of the European Union. When the international trade agreement known as GATT (General Agreement on Tariffs and Trade) was renegotiated recently, it was the Commission of the European Union, rather than European governments directly, which acted on behalf of Europe.

Fourth, the sovereignty of the state is challenged by international belief systems. Until recently this included communism, but the principal focus now is on the world religions. Many of the major world religions, including Judaism, Roman Catholicism and Islam, are able to exercise a degree of what is effectively political authority over at least some of their adherents, which is separate from, and implicitly a challenge to, the sovereignty of states.

ALTERNATIVE FUTURES FOR THE WELFARE STATE

Although the work of Offe and others points to the inherent crisis tendencies of the state, the evidence of actual crisis is ambiguous.[51] Welfare states have not, for example, been subject to complete collapse, and while in many cases they have been significantly restructured, governments continue to spend large proportions of their budgets on welfare and related activities. State welfare provision continues to enjoy significant levels of public support. Moreover, despite the coming to power in many Western countries of neo-liberal or neo-conservative parties, the proportions of state finance spent on welfare have proved remarkably resilient. What *has* changed are the ways in which those budgets are themselves divided up and the ways in which welfare services are delivered. The following future trajectories are not necessarily mutually exclusive (although some are) and nor are they all equally likely!

It seems likely that whatever other changes may occur, the blurring of the public, private and voluntary (non-profit) sectors will continue. Although in many countries it has never been possible to draw a completely rigid line between public and private welfare provision, there was a widespread acceptance until the 1980s that the public sector, if not inherently superior to the private sector, was essential to ensure that social benefits were distributed evenly and universally. Since then, however, both the commitment to universal provision (even where it was made most strongly) and the key role for the public sector have been increasingly challenged. Typically the state continues to be involved in the financing of service provision, at least for the most disadvantaged social groups, but the production of the services themselves is often undertaken by private or voluntary organizations under contract. This process of privatization has proceeded in a geographically uneven way both between and within different countries.

The growing importance of the informal and not-for-profit (or voluntary sectors) has led some commentators to talk about the emergence of a 'shadow state' – a set of institutions which are gradually taking over state functions partly as a result of political strategy on the part of the state, and partly through default.[52]

The second possible trajectory involves what has been called the 'hollowing-out' of the state. This process relates to the premise that the modern state is simultaneously too large and too small. It is too small to cope adequately with global economic forces, but too large to deal appropriately with the competing political demands and requirements of a heterogeneous population and regional mosaic. Hollowing-out supposedly involves a transfer of certain responsibilities upwards and outwards to supra-national organizations (of which the European Union is the only truly developed example) and a transfer of other responsibilities downwards and inwards to the grassroots, local or regional level.[53] This does not imply necessarily the loss of all power from the central state, however, which might still retain crucial functions in respect of citizenship allocation and the processes of social exclusion on which it is based.

One of the most developed accounts of the future of the state is to be found in the work of Bob Jessop. Jessop argues that the Keynesian welfare state is tending to be replaced with a new form of the state, the 'Schumpeterian workfare state'. Keynesianism was based on the government's control over the demand side of the economy. During downturns in the economic cycle, the government sought to boost demand, through tax cuts, for example, which, it was argued, would stimulate production to meet the increased demand and thus increase the level of employment. The key criteria for success from the Keynesian perspective is the preservation of full employment. By contrast, Jessop argues that in the globalized economy of the late twentieth century, governments are typically concerned with international economic competitiveness. For this reason, he suggests, government policy is concerned more with intervention on the supply side. For example, this might involve the development of new forms of education and training to provide a pool of skilled labour for potential investors, or reform of the institutions of the labour market to reduce protection for workers and increase labour market flexibility. The commitment in principle is to gaining a competitive advantage through innovation. Jessop suggests that such policies are 'Schumpeterian' after the economist Joseph Schumpeter, who was particularly concerned with the economics of innovation.

On the social side, Keynesianism was linked to a welfare state which sought to provide a minimum level of welfare for all, regardless of economic circumstances, and was thus based on a commitment (at least in principle) to social cohesion and integration. By contrast, Jessop argues that the new form abandons even a nominal commitment to social equality or minimum provision and seeks instead to use the social benefits system to socially engineer a more compliant and flexible workforce on the one hand while abdicating responsibility for a growing group of dispossessed 'have-nots'. The term 'workfare' is thus used not in the strict sense of 'food for work' programmes, but to indicate the underlying principle around which state social provision will increasingly be orientated. Central to the development of any such system would be the response of those at the bottom. Would the 30 per cent of the population who made up the 'have-nots' generate such social unrest to challenge the whole strategy? Or would the Schumpeterian workfare state become an authoritarian state capable of controlling and monitoring the dispossessed but unwilling or unable to provide them with a decent standard of living?

The final, and rather more optimistic scenario, is the one proposed recently in the case of Britain by the economic journalist Will Hutton.[54] Hutton argues that, far from being dead, Keynesian economics provides part of the answer to Britain's social and economic ills. Hutton identifies a resurgence of interest in Keynesian ideas. The neo-liberal, free-market approach which has dominated the thinking of many governments (including Britain's) in recent years, is based on an inaccurate picture of the economy. The free-market approach is based, Hutton points out, on the assumption of a 'level playing field' between all participants, in terms of information, market power and capacity to influence prices. Without such a level starting point, the operation of free markets can only produce markedly (and increasingly) uneven levels of economic and social welfare. Left to their own devices, markets do not tend towards equilibrium. Hutton argues that attention needs to be paid to reforming the institutions of the economy in tandem with constitutional reform of the state itself. Only then, he suggests, will the state have the capacity to manage and regulate the economy adequately and thereby to prevent the kinds of social dislocation foreseen above.

It is impossible to tell at this stage which of these various scenarios, if any, will develop to fruition. What is clear, however, is that while there are certainly processes at work, in the global economy and the institutions of the state, which will condition the direction of change, the development of welfare states will be, at least in part, the product of political struggles and strategies in the future, just as it was in the past.

Notes on Chapter 3

1 Unlike the state's territory, which, as we saw in Chapter 2, generally does.
2 Ron Johnston, 'The Rise and Decline of the Corporate-welfare State: A Comparative Analysis in Global Context', in Peter Taylor, ed., *Political Geography of the Twentieth Century: A Global Analysis* (London, Belhaven, 1993), p. 162.
3 Jean-François Bayart, *The State in Africa: The Politics of the Belly* (London, Longman, 1993). The book was first published in French in 1989.
4 Bob Jessop, 'The State as Political Strategy', in *State Theory: Putting Capitalist States in Their Place* (Cambridge, Polity Press, 1990), pp. 248–72. The essay was originally written in 1985.
5 *Op. cit.*, p. 249.
6 Note that 'boundaries' here does not (only) mean territorial boundaries, but also the boundaries between state organizations and other aspects of society.
7 Jessop, 'The State as Political Strategy', pp. 252–4.
8 *Op. cit.*, pp. 253–4.
9 *Op. cit.*, p. 254.
10 *Op. cit.*, p. 260.
11 For a further discussion of the concept of modernity, see Marshall Berman, *All That is Solid Melts into Air* (London, Verso, 1982).
12 John Gray, *Liberalism* (Milton Keynes, Open University Press, 1986), p. 82.
13 Christopher Pierson, *Beyond the Welfare State* (Cambridge, Polity Press, 1991), p. 110.
14 Ronald Johnston, Fred Shelley and Peter Taylor, eds, *Developments in Electoral Geography* (London, Routledge, 1990).
15 Gordon Clark and Michael Dear, *State Apparatus: Structures and Languages of Legitimacy* (Winchester, MA, Allen and Unwin, 1984); Peter Taylor, *Political Geography: World-economy, Nation-state and Locality* (London, Longman, 1989), pp. 204–50.
16 S. M. Lipset and Stein Rokkan, 'Cleavage Structures, Party Systems and Voter Alignments', in S. M. Lipset and Stein Rokkan, eds, *Party Systems and Voter Alignments: Cross-national Perspectives* (New York, Free Press, 1967). See also Stein Rokkan, *Citizens, Elections and Parties* (New York, McKay, 1970).
17 T. H. Marshall, *Citizenship and Social Class* (Cambridge, Cambridge University Press, 1950). His arguments have been reprinted in a more recent edition: T. H. Marshall, 'Citizenship and Social Class', in T. H. Marshall and Tom Bottomore, *Citizenship and Social Class* (London, Pluto Press, 1991), pp. 3–51.
18 This is clearly not always true. In South Africa under apartheid for example, the black African majority, as well as Asian and so-called 'coloured' minorities, were accorded only *de jure* civil citizenship and denied *de jure* political and social citizenship and any form of *de facto* citizenship.
19 Eleonore Kofman, 'Citizenship for Some but not for Others: Spaces of Citizenship in Contemporary Europe', *Political Geography* 14 (1995), pp. 121–37.

20 Stephanie Pincetl, 'Challenges to Citizenship: Latino Immigrants and Political Organizing in the Los Angeles Area', *Environment and Planning A* 26 (1994), pp. 895–914.

21 Carole Pateman, *The Disorder of Women: Democracy, Feminism and Political Theory* (Stanford, CA, Stanford University Press, 1989).

22 David Bell, 'Pleasure and Danger: The Paradoxical Spaces of Sexual Citizenship', *Political Geography* 14 (1995), pp. 139–53.

23 Philip Corrigan and Derek Sayer, *The Great Arch: English State Formation as Cultural Revolution* (Oxford, Blackwell, 1985), pp. 114–65; Christopher Dandekar, *Surveillance, Power and Modernity* (Cambridge, Polity Press, 1990), pp. 110–49; Miles Ogborn, 'Local Power and State Regulation in Nineteenth-Century Britain', *Transactions of the Institute of British Geographers* 17 (1992), pp. 215–26; Pierson, *Beyond*, pp. 6–39.

24 Pierson, *Beyond*, pp. 102–18.

25 Gøsta Esping-Andersen, *The Three Worlds of Welfare Capitalism* (Cambridge, Polity Press, 1990), pp. 26–7.

26 *Op. cit.*, p. 27.

27 *Op. cit.*, pp. 27–8.

28 David Harvey, *The Limits to Capital* (Oxford, Blackwell, 1982); Neil Smith, *Uneven Development* (Oxford, Blackwell, 1984).

29 Robert Bennett, *The Geography of Public Finance* (London, Methuen, 1980); Sarah Curtis, *The Geography of Public Welfare Provision* (London, Routledge, 1989); Steven Pinch, *Cities and Services* (London, Routledge and Kegan Paul, 1980).

30 David M. Smith, *Geography and Social Justice* (Oxford, Blackwell, 1994).

31 The 'crowding out' argument is particularly associated in Britain with the work of Roger Bacon and Walter Eltis. See Roger Bacon and Walter Eltis, *Britain's Economic Problem: Too Few Producers* (London, Macmillan, 1978).

32 J. C. Archer, 'Public Choice Paradigms in Political Geography', in A. D. Burnett and P. J. Taylor, eds, *Political Studies from Spatial Perspectives* (New York, John Wiley, 1981).

33 C. M. Tiebout, 'A Pure Theory of Local Expenditures', *Journal of Political Economy* 64 (1956), pp. 416–23.

34 Leslie Hepple, 'Destroying Local Leviathans and Designing Landscapes of Liberty? Public Choice Theory and the Poll Tax', *Transactions of the Institute of British Geographers* 14 (1989), pp. 387–99. The decentralizing force of the measure was quickly undermined when it became clear that far from resulting in general downward pressure on local taxation and spending, local governments and their electors were in many cases opting for a tax-service mixture which did not fit in with the central government's strategy of constraining public expenditure and curbing the expansion of local government.

35 Cynthia Cockburn, *The Local State* (London, Pluto Press, 1977).

36 On the US, see: Dennis R. Judd and Todd Swanstrom, *City Politics: Private Power and Public Policy* (New York, HarperCollins, 1994); and Michael Peter Smith, *City, State, and Market: The Political Economy of Urban Society* (Oxford, Blackwell, 1988). On the UK, see: Allan Cochrane, *Whatever Happened to Local Government?* (Buckingham, Open University Press, 1993); Simon Duncan and Mark Goodwin, *The Local State and Uneven Development* (Cambridge, Polity Press, 1988); and David Wilson and Chris Game, *Local Government in the United Kingdom* (Basingstoke, Macmillan, 1994).

37 Gerry Stoker, *The Politics of Local Government* (Basingstoke, Macmillan, 1988).

38 Clark and Dear, *State Apparatus*, pp. 131–52.

39 The heterogeneity of political language and discourse is stressed by Clark and

Dear, although they arguably over-estimate the capacity of the state to control and structure language and under-emphasize the language of resistance: Clark and Dear, *State Apparatus*, pp. 81–103. I have provided an brief discussion of discursive conflicts *within* local state institutions in Joe Painter, 'The Culture of Competition', *Public Policy and Administration*, 7,1 (1992), pp. 58–68.

40 Will Hutton, *The State We're In* (London, Cape, 1995); Geoffrey Ingham, *Capitalism Divided: The City and Industry in British Social Development* (Basingstoke, Macmillan, 1984).

41 Parminder Bakshi, Mark Goodwin, Joe Painter and Alan Southern, 'Gender, Race and Class in the Local Welfare State', *Environment and Planning A* (forthcoming).

42 Jürgen Habermas, *Legitimation Crisis* (Cambridge, Polity Press, 1988), first published in German in 1973; James O'Connor, *The Fiscal Crisis of the State* (New York, St Martin's Press, 1973); James O'Connor, *The Meaning of Crisis* (Oxford, Blackwell, 1987); Claus Offe, *Contradictions of the Welfare State* (London, Hutchinson, 1984).

43 Offe, *Contradictions*, p. 58.

44 Christopher Pierson provides further details in relation to each of these critiques. See Pierson, *Beyond*. Other useful material is in: Roger Burrows and Brian Loader, eds, *Towards a Post-Fordist Welfare State?* (London, Routledge, 1994); Glenn Drover and Patrick Kerans, eds, *New Approaches to Welfare Theory* (Aldershot, Edward Elgar, 1993); Jane Lewis, ed., *Women and Social Policies in Europe: Work, Family and the State* (Aldershot, Edward Elgar, 1993); Fiona Williams, *Social Policy: A Critical Introduction* (Cambridge, Polity Press, 1989).

45 For a useful survey of feminist political thinking, and an outline of the main aspects and history of each of the three strands, see Valerie Bryson, *Feminist Political Thought: An Introduction* (Basingstoke, Macmillan, 1992).

46 William Beveridge, quoted in J. Clarke, A. Cochrane and C. Smart, *Ideologies of Welfare* (London, Hutchinson, 1987), p. 101.

47 See, for example, Mike Savage and Anne Witz, eds, *Gender and Bureaucracy* (Oxford, Blackwell, 1992).

48 Charles Tilly, *Coercion, Capital and European States* (Oxford, Blackwell, 1990), pp. 62–6.

49 For details, see Pierson, *Beyond*, pp. 79–95.

50 The main exception is anarchism, which seeks to develop a society without a state.

51 For a review, see Pierson, *Beyond*, pp. 140–78.

52 Jennifer Wolch, 'The Shadow State: Transformations in the Voluntary Sector', in Jennifer Wolch and Michael Dear, eds, *The Power of Geography* (Boston, MA, Unwin Hyman, 1989), pp. 197–221.

53 Bob Jessop, 'The Transition to Post-Fordism and the Schumpeterian Workfare State', in Burrows and Loader, eds, *Towards*, p. 24.

54 Hutton, *The State We're In*.

|4|

Imperialism and post-colonialism

Overview

In this chapter I want to turn to another aspect of the geography of state formation: the relationship between those first modern states of Europe and other parts of the world. Paralleling their formation through warfare, the connections of modern states with other places and peoples have often been violent and bloody. 'Imperialism' is the control by one state of other territories. 'Colonialism' refers to the establishment of permanent or extended settlement (colonies) in those territories. Imperialism may be military and political (direct, or formal, imperialism) in which the government of the territory concerned is taken over by the imperial power, or it may be economic (indirect, or informal, imperialism) in which the territory is formally independent but tied to the imperial power by (unequal) trading relations. In addition, it is now becoming increasingly common to identify cultural imperialism, in which existing or traditional ways of life and ways of thinking are subordinated to the culture of the imperialists.

The geographies of imperialism are geographies of political strategies of suppression, domination and exploitation; they are also geographies of resistance which continue to develop and change into the present day. I will begin by briefly outlining the development of European imperialism and then consider some of its formative links with academic geography. I will then turn to the political strategies of colonizers and colonized and see how these continue today with the development of post-colonialism.

The expansion of Europe

Encounters with 'other' peoples

The reasons why a small number of small European states were able to
dominate and control such a huge proportion of the globe, in terms both of
land area and people, are hotly disputed.[1] A range of explanations have been
advanced, including relative levels of technological development in Europe
and other places, political conflict between European states, the emergence
of capitalism, Western Europe's maritime traditions and expertise, and so
on. Each of these factors no doubt played their part, and it seems increasingly unconvincing to rely on a mono-causal explanation. What is also
clear, though, is that there was nothing inevitable about European geopolitical superiority. Wherever the Europeans went they found other people,
who were, more often than not, living in complex societies with high levels
of technological sophistication, political organization and cultural development. One of the mistakes of early imperialists was to attempt to judge the
cultures and peoples they encountered by European standards, and thereby
to fail to recognize that in their own terms they were as sophisticated and
'advanced' as the supposedly superior European societies. The fact that
these 'other' peoples did not set out to rule and dominate the rest of the
world was not a product of their so-called 'primitive' condition or 'degenerate' social structures, but rather reflected very different combinations of historical circumstances and political, cultural and economic priorities and
values.

Although the modern world, and particularly the relations between the
rich industrialized countries of the North and the poorer countries of the
South cannot be understood without the context of Western imperialism, it
would be a mistake to give the impression that European control was ubiq-
uitous, complete, or homogeneous. Parts of the globe escaped formal
European control altogether, but even places which were part of formal
European empires were never wholly subordinated, or subdued. First, there
was the logistical problem of governing tracts of territory and populations
which were larger than the compact countries of Europe and often a long
way away. The practical difficulties of colonial administration meant that in
many cases the imperial powers had to incorporate and buy off local political leaders, and, to some extent, adjust to local political and social structures.
To say this is not intended to lessen the horrors of the practices of imperialism, but to recognize the force of the points made in Chapter 1, that any sustained system of rule or governance depends on coalition and alliance
formation, and the use of the resources that are to hand. Imperial rule was
constituted through active *engagements* between the strategies and institutions of rulers and ruled, albeit ones which were highly unequal and unjust.

Second, imperialism was always resisted. Resistance did not always take the form of organized political struggle (although it often did), but included everything from sabotage and military actions, through civil disobedience and strikes to minor non-co-operation, foot-dragging and grumbling. Not all of these strategies were equally successful, of course (nor were they all equally strategic). But wherever Europeans went and tried to govern, people resisted them.

The motives for expansion

If the underlying reasons for the relative success of European domination were complex, so too were the motives which prompted European expansion. Certainly, no account of imperialism can ignore trade. Mercantile capitalism, which was the mode of economic organization of Europe's cities in the later Middle Ages, was based on a simple principle: buy cheap and sell dear. Many goods could be produced within Europe from local raw materials, including linen and woollen cloth, wood and timber, leather goods, a variety of foodstuffs, and wine and beer.

With the growth of medieval cities and the consequent development of a market for luxury products there was an increasing demand for raw materials and goods which could not be produced at home, or which were in short supply. These included silk and cotton, spices, and precious metals and stones. Good sources of many luxuries in Asia were known to Europeans and by 1400, as Europe teetered on the brink of modernity, there were already long-established overland trade routes to the East. However, the land routes were insecure and subject to delays, loss of cargo and the whims of rulers along the routes:

> Before the voyages of discovery and the opening of the oceans to trade, the Italian city-states, especially Genoa and Venice, grew wealthy . . . on the strength of their strategic position between Europe and western terminuses of the Asian trade routes. Vigorously commercial and active developers of maritime technology, they were frequently at war with one another in the competition for shares of the Eurasian trade. Initially Islam had been a boon to these merchants because it inhibited Muslim merchants in the Levant from venturing into Christian territories. But when the Turks captured Constantinople in 1453, thus making the Byzantine capital a Muslim city, Genoa and Venice lost an important outpost in what was becoming an increasingly hostile world. The city-states were soon to enter a long period of decline as the maritime strength of the west passed first to southwest and then to northwest Europe.[2]

The development of an all-water route to Asia became a priority. Because he first encountered the lands we know today as America, Christopher Columbus is probably the best known of the merchant adventurers and sea-farers of the so-called 'age of discovery'. However, his voyages westwards in 1492, 1493, 1498 and 1502 were less immediately significant commercially than that of Vasco da Gama, who in 1497 travelled south past the huge con-tinent of Africa and then east reaching the south-western coast of India on 22 May 1498.[3] For the first time it was possible for Western Europe to trade with Asia without risking the difficult overland routes through the Near East.

The other great motive for overseas expansion was religious. Early explorations by Spain and Portugal were impelled in part by the perceived threats to Catholic Christendom from Islam in the east and the Protestant Reformation to the north. Wherever Iberian explorers went they claimed land for their monarchs, but also souls for the Church. Later, in the seventeenth century it was Protestantism which sought salvation overseas, with the settlement of the eastern seaboard of North America by the Puritans.

Iberia abroad

The new maritime trade routes to the East were dominated initially by Portugal, whose seafarers established trading posts around the coasts of Africa and South Asia and in the Far East. The Portuguese emphasis on trade and commercial activity meant that to begin with the Portuguese empire in Africa and the East was made up of many tiny possessions. Little attempt was made to acquire large areas of territory in the hinterlands of the trading posts. In due course the significance of Columbus's encounter became evident and the first extensive overseas European empires were those established in the New World by Spain and Portugal. The Spanish empire in particular expanded rapidly through the Caribbean, Central and South America and north through Mexico into present-day California, Arizona, New Mexico, Texas and beyond. The Portuguese were also active in South America after Cabral's landing in 1500, expanding into present-day Brazil.

The Iberian expansion generated vast wealth for the Spanish and Portuguese crowns which were also united from 1580 to 1640. Precious metals and particularly silver were the sources of wealth, which was stripped from America by the tonne and shipped back to Europe. In the process the conquistadors, the soldier-adventurers responsible for the process of colonization, laid waste to great civilizations, such as the Mayas, the Incas and the Aztecs. This occurred both through military conquest and as a result of the arrival of European diseases which were previously unknown, and to which the local people thus had little resistance.[4]

The beginnings of the British empire

Within only a few decades, however, the dominance of Spain and Portugal was under threat. By the second half of the sixteenth century, the British and French exploration of the North and east of the North American continent was well under way. British, French and Dutch colonies were established along the Atlantic seaboard, British explorers sought in vain for a north-west route to Asia which would bypass the Spanish, while the French were busy opening up the interior in Canada and eventually travelling down the Mississippi to the Gulf of Mexico to what became Louisiana. The French influence is still in evidence in the French place-names along the great river and the local language of Louisiana and New Orleans.

Meanwhile, in the East, the Dutch (who gained independence from Spain in 1584) and the British were in gradual ascendance, challenging the Portuguese monopoly of trade. In 1599, the British East India Company was founded, and chartered by Queen Elizabeth in the following year. It was followed in 1602 by its Dutch counterpart. Contact between Europe and the East during the seventeenth century was largely a commercial affair. The Dutch headed for south-east Asia and the spice islands, while the English struggled to develop trading links with the Mughal emperor in India. Gradually the English gained in strength relative to the Portuguese who had long-established trading posts in India. However, the East India Company was by no means equivalent to the military conquerors of South and Central America. In the Mughal empire they met a civilization which, on land, was significantly stronger and which expected to receive tribute and deference from these overseas visitors.

Though stronger than the tiny forces of the East India Company, the Mughal empire was coming to an end. During the first half of the eighteenth century it collapsed and a variety of would-be successors vied for supremacy. Both the British and the French sought to capitalize on the confusion, with the British eventually prevailing. When Robert Clive gained political power in Bengal in 1757, he unleashed a tide of greed and exploitation in which the merchants of the East India Company plundered the local economy, destroying the sources of their wealth and bringing the Company to the verge of bankruptcy. This led in 1773 to action by the British Government in London to bring the Company under state regulation. Warren Hastings was appointed as first Governor-General of all the British possessions in India. As the British were losing their North American colonies in the War of Independence (1775–83), they were securing their hold over what was to become known as the 'jewel in the imperial crown'. However, although hegemonic throughout the sub-continent, the British came to control only about two-thirds of the land area directly, with the remainder under the nominal rule of local princes. Moreover, the process of empire-building was a long and slow one taking a good century to complete, and involving military, economic and cultural struggles with local people and institutions. Over time the emphasis of the British activities in India shifted from trading as equals, or even subordinates, through the extension of trade through political and military means, to an assumption that the British were in India to rule, with trade left to private individuals and companies. This process culminated with the Indian revolution of 1857, the resulting transfer of powers from the East India Company to the Crown in 1858 and the naming of Queen Victoria as Empress of India.

At this point, however, the final wave of imperialist expansion had hardly begun. Africa was a vital and bloody link in the 'triangular trade' which brought Africans to the Caribbean and the American South to be sold as slaves to work on the cotton, sugar and tobacco plantations. These raw materials were then exported to Europe and manufactured into goods for re-export to the colonies. All around the coast of Africa Europeans had established small colonies and trading ports. In the early nineteenth century the interior of the continent was almost completely unknown to Europeans. However, in the 34 years between 1880 and the First World War, the entire continent, its people and resources, had been carved up between the European powers.

Finally in the South Pacific, the search for the Great Southern Continent, which had been a presence in European mythology during the Middle Ages, was finally discovered to be an island following its circumnavigation in 1801–2. Until 1851 the British Government used Australia as a penal colony, and for the next 50 years, particularly with the discovery of gold, it was the focus of considerable emigration, becoming, with New Zealand, a major exporter of agricultural products. Neither of the two was an empty land, however, and as in other parts of the globe, local people found their cultures and livelihoods altered and often destroyed by the colonizing Europeans.

Geography's imperial roots

Imperialism and the role of geographical knowledge

In this process of imperial expansion and overseas colonization, knowledge was of vital importance. Not only did European expansion produce new knowledge at an ever-increasing rate, but it also depended upon knowledge. This dependence was practical in the sense that the development of overseas possessions required specific kinds of information and skills, such as cartography, surveying, ship-building, astronomy and navigation, settlement construction, mining, agriculture and so on. It was also a dependence on particular ways of knowing *about* other places and people. The arrogance with which Europeans asserted their dominion over the globe and the ease with which they could bring themselves to destroy, kill, and maim other people and disparage their cultures and achievements was based on certain preconditions. In order to be able to carry out these acts, it was necessary for Europeans to understand themselves as superior and others as inferior. The barbarities of colonialism rested on a set of assumptions, representations and discourses about the rights of Europeans in relation to the rest of the world. We will consider some of these discursive practices in the following section. First, though, I want to look briefly at the role played by geography in these processes.

The modern discipline of geography was, perhaps more than any other, the product of imperialism. First, European knowledge of the earth's surface – its land masses and seas, its plants and animals, its peoples and their ways of life – was gained very largely through the process of European expansion and formed the content of the emerging subject of geography. Second, many of the practical skills and knowledges through which exploration and settlement were practised are central to geography, from map-making to settlement planning. Third, geography operated through particular ways of knowing the world which both enabled and legitimated the practice of imperialism.

As I mentioned in Chapter 1, the origins of the sub-discipline of *political* geography were entangled with imperialist rivalries between the European powers. Such entanglement was not limited, however, to just one sub-discipline of geography. As David Livingstone, and many others, have pointed out, the development of the academic subject of geography as whole was both the product of, and implicated in, the expansionist policies of Europe throughout the age of empire.[5] One clue is provided by timing. As David Stoddart avers, 'what distinguishes geography as an intellectual activity from ... other branches of knowledge is a set of attitudes, methods, techniques and questions, all of them developed in Europe *towards the end of the eighteenth century.*'[6]

For Stoddart the new methods and techniques which gave modern geography its coherence were united by their stress on direct observation[7] and what he regards as a move away from interpretations of the world based on myths, legends and fantastic narratives. Other commentators, however, have suggested that the writings and works of modern geography, particularly in its nineteenth-century phase were rather less than objective, and were saturated with European assumptions about the world and its contents. The narratives may have been different, but they were stories none the less and closely connected to moral and cultural views about the relationship between the 'West' and its 'Others'.

The discourse of climate

David Livingstone, for example, places considerable stress on the relationships between geographers' interpretations of climate and climatic zones, and the discourses of racial inferiority and superiority which were central to the imperial project. According to Livingstone, geographers' study of climate was far from the objective science which it was claimed to be. Rather, notions about climate became entangled with moral, religious and political judgements about people. The impact of climatic variation on humanity was thought to be highly significant and to condition not just agricultural production, but whole ways of life and people's biological make-up. What Livingstone calls the 'moral economy' of climate operated through linking climatic variation closely to the supposed division of the human species into different races.

Nowadays social scientists are much more critical of the whole idea of race as a biological distinction. Indeed many argue that there are no satisfactory biological grounds for grouping people into races at all and there are certainly no inherent differences between the potential physical, mental and emotional characteristics of members of supposedly different 'races'. For many decades, however, it was widely, almost universally, assumed that the human species consisted of biologically distinct groups ('Caucasian', 'Negro', 'Asiatic' and so on). Indeed some writers even believed that Africans were actually a separate species and represented an earlier evolutionary stage in the development of modern *homo sapiens*.

The study of climate was significant, because it was argued that different races suited different types of climate, either because racial differences were produced by climatic ones, or because the different races had been allocated 'by nature or God [to] climatically appropriate regimes'.[8] Thus it was thought that the bracing climates of north-west Europe were responsible for producing a race of hard-working, intelligent and rational people, the warm Mediterranean climate generated the relaxed and emotional 'Latin temperament', while the hot climates of the tropics led to the moral and physical degeneracy and indolence thought to be characteristic of the Africans.

It is not possible to dismiss these ideas either as simple, if barbaric, racism, or as immature and underdeveloped science, for their significance was widespread. Not only did the discourse of 'climate's moral economy' underpin and provide scientific justification for what became the routine practices of nineteenth-century imperialism (such as slavery), it also exerted a surprisingly long-lived influence on geography as an academic discipline:

The idea that climate had stamped its indelible mark on racial constitution, not just physiologically, but psychologically and morally, was a motif that was both deep and lasting in English-speaking geography. ... in Austin Miller's standard textbook on *Climatology*, first published in 1931, he explained that 'Psychologically, each climate tends to have its own mentality, innate in its inhabitants and grafted on its immigrants ... the enervating monotonous climates of much of the tropical zone, together with the abundant and easily obtained food-supply, produce a lazy and indolent people, indisposed to labour for hire and therefore in the past subjected to coercion culminating in slavery.' What is remarkable here is the way moralistic terms – enervating, monotonous, lazy, indolent – were still presented as settled scientific maxims.[9]

Even as late as 1957, the following words of the renowned geographer Griffith Taylor were being republished: 'The writer believes that it is precisely because the Negro was thrust into the stagnant environment of the Tropics ... that he preserves so many primitive features ... racial evolution' has left the Negro 'far behind'.[10]

Mapping and naming

As a subject, geography was also implicated in the imperial project in highly practical ways. Controlling and ruling distant lands and peoples required knowledge of the land as much as the people. Maps and charts were central to European strategies throughout the overseas empires. Mapping brought the land into sight and into Western frameworks of understanding. It allowed order and Western rationality to be imposed on human landscapes shaped through very different world views. In *The Road to Botany Bay*, Paul Carter discusses the significance of the survey and the map for colonists in Australia:

Maps were understood as ways of getting in. ... Exploring and surveying were ... two dimensions of a single strategy for possessing the country. The map was an instrument of interrogation, a form of spatial interview which made nature answer the invader's need for information.[11]

European colonialism also sought to possess the land through naming and labelling it either in familiar terms with English, French or Spanish words, or by using the words of local people. In either case, as Carter points out, the act of naming, of inscribing maps, and thus the land linguistically, was another strategy through which the land could be made knowable, and thereby possessed. In Australia, the historical space of the white settlers emerged through the medium of language. But the language that brought it into cultural circulation was not the language of the dictionary: on the contrary it was the language of naming, the language of travelling. What was named was not something out there; rather it represented a mental orientation, an intention to travel. Naming words were forms of spatial punctuation, transforming space into an object of knowledge, something that could be explored and read.[12]

The close links between geography as a system of knowledge, geography as a practical activity and the expansion of European empires have been made increasingly clear in recent years.[13] An interest in imperialism among geographers continues to this day, albeit usually in a rather more critical vein. Until recently, this interest was largely in the political economy of imperialism. Although that perspective continues, contemporary writers have focused increasingly on imperialism as a way of thinking about and looking at the world, of constructing the identities of ourselves and of others, and of attempting to control not only the economic and political fates of other peoples and lands, but their cultural destinies as well. In this geographers have increasingly drawn on the writings and ideas of the so-called 'post-colonial' theorists such as Edward Said and Gayatri Chakravorty Spivak. I will return to these more recent attempts to write more critical geographies of imperialism later in the chapter. Now, though, I want to turn to the development of imperialism itself and examine its spatial practices in a little more detail, beginning with a discussion of one particular theory of the expansion of Europe overseas: world-systems theory.

World-systems theory

An outline

Interpreting and explaining the broad sweep of the expansion of Europe into the rest of the world is a complicated business, and has been the focus of much debate and academic dissent. One framework is 'world-systems analysis' which has been developed over many years by Immanuel Wallerstein. It is a perspective which has been particularly influential within political geography through the work of Peter Taylor, and especially his *Political Geography: World-economy, Nation-state and Locality*. First published in 1985 (and part of an upsurge of interest in political geography) this book was one of the first, and is probably still the best known of, attempts to place the field of study of the sub-discipline of political geography comprehensively within a coherent theoretical framework. At the time of writing it has already reached its third edition.

Wallerstein draws on the pioneering work of economic historians Fernand Braudel and Karl Polányi. Central to Braudel's ideas was the concept of *longue durée*, which places stress on the long timescale of social and economic change over decades or even centuries. Polányi proposed that, historically, economic activity has always taken one of three forms. The first of these is the *reciprocal-lineage* system, which is characteristic of the traditional societies within which human life has been organized for most of its history. In this system, exchange between producers takes place on a mutual basis (hence 'reciprocal') and is organized through kinship groupings (hence 'lineage'). The second form is the *redistributive-tributary* form, in which there is a net 'upwards' redistribution of the products of human labour from producers to a dominant group, such as occurred during feudalism. The third form is *market exchange* in which goods are exchanged 'freely' in a market. This form is typical of capitalism. Although in principle these forms can exist together in time and space, Polányi argues that one is likely to dominate the others and thus structure the overall character of the economy. In addition, over time, there has been a move from the reciprocal form through the redistributive form to the market form.

Wallerstein shares Braudel's stress on long-term shifts in economic and social relations. He argues that Polányi's three forms of exchange correspond to three distinctive types of social system; indeed these three types are the only kinds of socio-economic system which have existed. They are *mini-systems*, in which exchange is reciprocal; *world-empires*, in which exchange is redistributive; and a *capitalist world-economy* in which market exchange dominates. Historically, mini-systems have been by far the most numerous, although in the modern world we know of only a tiny proportion of them and today there are probably no remaining cases. They include, for example, the social systems of certain North American tribes prior to the expansion of European settlement. According to Wallerstein, there have been many world-empires, which have a large base of agricultural producers providing both subsistence for everyone and luxuries for a small élite group. They include the Roman Empire, the Chinese Empire, and the feudal system in medieval Europe. According to world-systems analysis, all mini-systems and world-empires have been eliminated or incorporated into the only remaining system: the capitalist world-economy. There have been previous examples of nascent world-economies, based on market exchange, but until the sixteenth century they were quickly incorporated into existing world-empires. From the sixteenth century onwards, however, the world came to be dominated by the European world-economy, which, Taylor suggests, became truly global in about 1900.

Following Taylor,[14] we can identify two important insights from Wallerstein which distinguish the world-systems approach from traditional conceptions of global economic change. The first of these is the 'one-society assumption'. Traditional social science has assumed that the world consists of many 'societies' which are normally seen as being the same as countries. Hence we speak of British, American or Chinese society. This, says Wallerstein, is mistaken. The integration of economic activity in the world-economy means that there is now just one, global, society. This insight is related to the second, which Taylor calls the 'error of developmentalism'. Traditionally, 'development' has been seen as a path along which the multiple societies (countries) pass from low levels of economic activity to more complex and wealthy ones. Because there is only one world-economy, it is not possible for individual bits of it to pass independently up what Taylor refers to as a 'ladder' of development. The economic activities going on in all countries are closely related to each other. The capacity of some countries to produce a great deal and to sustain high material standards of living depends upon the existence of other countries whose economies are actively 'underdeveloped' by the processes of the world-economy to promote wealth for the few at the top. These two insights are very important, particularly as correctives to other, more conventional, assumptions about the relations between societies.

Wallerstein's approach, therefore, offers a broad framework within which the expansion of Europe, which I described above, may be understood. There are clear parallels with the stress I have placed on the emergence and development of the overseas empires as a historical process. The world-systems approach has attracted both fervent support and criticism from academics and, as I have mentioned, has had an important influence within political geography. Its existing stress on the *spatial structure* of the world-economy, which Wallerstein argues is divided into a core, a semi-periphery and a periphery is highly geographical; this aspect of world-systems theory has been carefully developed by Peter Taylor. Despite its attractive elements, however, it is not in the end compatible with the approach adopted in this book.

Critiques of the world-systems approach

There are a number of principal criticisms which have been made of Wallerstein's ideas. Here I will rely on Anthony Giddens's summary of them.[15] Giddens suggests that world-systems theory is flawed in two important ways. First, he argues, it suffers from economic reductionism. This does not mean that it only looks at economic processes, but it does imply that where politics and culture are examined, they tend to be explained in terms of the economy. Thus, the dynamic of state formation, which I examined in Chapter 2, and which seemed to be closely shaped by war and military strategy, is considered in Wallerstein's account as a feature of the development of the world-economy. In Giddens's view, and mine, the dynamics of the world-economy are crucial in explaining the changing world, but that changing world is also a product of the development of the international system of states, which cannot be accounted for entirely in economic terms. This also means that the one-society assumption in Wallerstein's approach needs to be modified. There may be only one capitalist world-economy, but British 'society', French 'society' and American 'society' do have meaning in other ways. Since each is a territorial state which accords, for example, citizenship to its population in different ways, being part of British 'society' is politically very different from being part of French 'society'. The same point could be made in terms of religious, national, linguistic or ethnic groupings, although these would not map neatly on to modern states in the way that citizenship does. In other words the one-society perspective holds more true when 'society' is being thought of as a system of *economic* integration, and does not work so well when political or cultural relations are being considered.

The second difficulty that Giddens encounters with Wallerstein's ideas is the *functionalist* element within them. Functionalism, in the sense Giddens uses it, means explaining something in terms of its effects. This is common in the biological sciences. For example, we explain that fish have gills because the effect of possessing gills is to enable 'breathing' underwater. In thinking about social systems, an interpretation is functionalist if it explains one feature of a system in terms of the function it fulfils in helping to keep the system as a whole going. Giddens argues that Wallerstein's category of 'semi-periphery' is an example of his functionalist thinking, because 'the existence of semi-peripheral regions is explained by reference to the "needs" of the world system'.[16] From Giddens's perspective, social systems cannot have 'needs', only people can. Now, it may be true, as Wallerstein says, that the existence of a middle tier of semi-peripheral countries between the rich core and poor periphery helps to stabilize the world-economy. But this stabilizing function cannot account for the initial emergence, or the continuing existence of the semi-periphery.[17]

Strategies of colonial domination

The interpretative framework I outlined in Chapter 1 seeks to avoid both economic reductionism and functionalism. First, it avoids functionalism by stressing politics as the pursuit of strategies (whose outcomes are uncertain and which may be disruptive rather than functional to the integration of the social system). Thus investigating (and explaining) the pattern of imperialist expansion involves studying the strategies of both colonizers and colonized in their particular context. This also means that the integration of the world beyond Europe into the world-economy was (and remains) rather less complete and comprehensive than Wallerstein's account suggests. Second, it avoids economic reductionism by emphasizing that political strategies, and the resources on which political power depends, are not just economic but also cultural, military, patriarchal, racist and so on. These other forms of power are qualitatively distinct from capitalist economic power and have different histories, geographies, preconditions and effects. In other words, imperialism was as much about strategies of cultural domination of the rest of the world as it was about strategies of economic exploitation and control.

Economic and military dimensions

The economic and military strategies through which Europe's overseas empires were put in place have been widely studied. In general the opening up of territory and the implementation of imperial government in Europe's overseas colonies was carried on by military, or quasi-military, means. Military strategies varied widely between different imperial powers. In the early phase of Iberian expansion in South America, the military campaigns were of crucial importance in wiping out existing forms of social and political organization. By contrast, in south Asia, the British exploited the *political* weakness of the Mughal state, and, while military activity was important at various times, the economic and administrative incorporation of local élites was also of crucial significance.

One of the standard accounts of European imperialism is provided by the historian D. K. Fieldhouse. Although it is clear that Europe gained massively in economic terms during the early phases of overseas expansion, Fieldhouse argues that Europe's later (nineteenth- and early twentieth-century) overseas empires were not subject to economic 'exploitation' by the imperial powers. While I do not entirely share this view, Fieldhouse is correct in suggesting that the later empires in tropical Africa and Asia were certainly not established initially for the purpose of economic profit-making:

> Modern empires were not artificially constructed economic machines. The second expansion of Europe was a complex historical process in which political, social and emotional forces in Europe and on the periphery were more influential than calculated imperialism. Individual colonies might serve an economic purpose; collectively no empire had any definable function, economic or otherwise. Empires represented only a particular phase in the ever-changing relationship of Europe with the rest of the world: analogies with industrial systems or investment in real estate were simply misleading.[18]

As he goes on to point out, however,

though the colonial empires were undoubtedly functionless in origin, this is not to say that they did not later provide an economic return, a 'profit', to their owners. Certainly many colonial enthusiasts alleged that they could and did.[19]

Despite his doubts about the exploitative character of later imperial rule, Fieldhouse identifies six ways in which economic advantages accrued from it and these usefully summarize the economic strategies involved.[20]

1 'Looting' an occupied territory of its treasures. This occurred, for example, with the shipment of precious metals from Mexico and Peru.
2 The transfer of colonial revenues to the metropolitan treasury.
3 The transfer of money to the imperial metropolis in the form of 'interest on loans, payment for services rendered, the pensions and savings of colonial officials and the profits made by business firms'.[21]
4 The imposition of unequal terms of trade on a colony.
5 The exploitation of natural resources without corresponding compensation.
6 The availability of higher rates of return on investments in the colonies than on investments at home.

This last aspect of imperial economics has been of particular significance because it was the focus of one of the earliest and most influential interpretations of imperialism: that of the Russian revolutionary leader, Vladimir Ilyich Lenin.[22] Lenin defined imperialism in terms of a number of characteristics:

1 capital is exported from the imperial economies (instead of only finished manufactured goods);
2 production becomes concentrated in the hands of a group of large companies;
3 banking capital is merged with finance capital;
4 the world is divided between the imperial (capitalist) states;
5 imperialist expansion buys off social dissent at 'home'.

This formulation is interesting in the context of my framework because of the stress placed on strategic behaviour by state governments and capitalist enterprises. Imperialism is seen as a strategic response to particular problems faced by business in the wealthy 'core' countries. According to Lenin, these included the lack of markets for the output from increased production as well as social and political discontent.

According to Fieldhouse, the evidence relating to the existence or otherwise of these kinds of relationships is ambiguous. It is clear that the earlier period of imperialism, during the sixteenth, seventeenth and eighteenth centuries was much more significant economically than the later, nineteenth- and twentieth-century phase. Where profit-taking is in evidence, it is not clear, Fieldhouse suggests, whether it occurred because of imperialist government or in spite of it. In terms of political strategies, however, it is clear that overseas empires were supported by many politicians and industrialists at home because of their perceived economic returns. It is also clear, however, that by the time formal imperialism reached its high point at the start of the twentieth century, the maintenance of elaborate colonial governments, administrations and security forces in Africa and Asia was rapidly becoming an economic drain on the European powers. This suggests that nineteenth-century expansion and the sustaining of imperial control involved a range of strategies other than purely economic ones.

Cultural and discursive dimensions

By contrast, cultural and discursive strategies of imperial domination have not been examined in the same detail until recently. This does not mean, however, that they were of any less consequence than those of a military or economic nature, not least because in many respects they continue to this day with the widespread stereotyping of the Third World as degenerate, corrupt, incompetent and violent. Even the more humanitarian approaches to materially poor countries (such as those associated with some overseas aid) have the potential to be quite patronizing, as is exemplified in the often-used slogan, 'helping them to help themselves'.

Discursive strategies are important because they embody particular understandings of the respective roles and natures of European and colonized people.[23] Such understandings can be necessary preconditions for military or economic exploitation. Thus it is considerably easier to kill someone if you have been brought up to see them as less than fully human, or as a member of an 'inferior' race. It may be easier to reorganize a system of production to appropriate profits 'back home' if local forms of industry are represented as less efficient or more primitive, regardless of their actual productivity.

However, the cultural relationship between 'the West and the rest'[24] was rather more ambiguous than this. It was not the case that the Europeans merely regarded themselves as 'superior' to other 'races'. The discursive strategies of imperialism depended on constructing 'the rest' as qualitatively different from 'the West', not only as inferior. One aspect of these strategies involved a discourse which contrasted a familiar, everyday 'West' with an exotic 'Other'. Among other things this involved sexual exoticism. For example, the 'Orient' is often represented as sexually degenerate, or the setting for exciting and exotic sexual encounters.

Anton Gill has charted the complex and ambiguous sexual discourses and practices of empire in his book *Ruling Passions*, written to accompany a BBC television series of the same name.[25] He writes of India:

> allure and repulsion sometimes went hand in hand. In the less hide-bound eighteenth century, allure had the upper hand, and it main-tained its supremacy for many – perhaps one could say for most – as long as the British were in India. For some it was a simple proposition. Edward Sellon, a British Indian Army officer of the 1840s, had no embarrassment about the joys of the east: 'I now commenced a regu-lar course of fucking with native women. They understand in perfec-tion all the arts and wiles of love, are capable of gratifying any tastes, and in face and figure they are unsurpassed by any women in the world. . . . It is impossible to describe the enjoyment I have had in the arms of these syrens . . .'

In these discourses Western men were presented as the very embodiment of virile, upright manhood. In the early 1920s a small book appeared called *The Romance of Empire*. Aimed at British schoolboys it seems to have been popular, as the edition I found in a second-hand bookshop shows that 24,000 copies had been printed. The book opens with a preface in which the author, Philip Gibbs, claims that the 'making of the British Empire has been a great adventure of which we may well be proud'. It was, he contin-ues,

> an adventure in which the manhood of the race has proved its met-tle, time and time again, through many centuries and in many lands; an adventure in which men have spilt their blood freely, with a genial courage, with a really rollicking spirit of gallantry, and with a fine carelessness of danger and death. . . . Not yet has the time come when the audacity of a brave man in a tight place, the steady nerve of a strong man in a dangerous encounter, the quick wit of a gallant fellow in a difficult enterprise, shall not be hon-oured and admired.[26]

By contrast, the world beyond Europe was often presented in feminized forms. Early images depicting the initial encounters between the West and the New World, for example, often showed America as a woman. For the West, which prided itself on its Enlightenment masculine reason, this sym-bolism not only made the New World seem inferior, socially and culturally to the West, but also served to emphasize the exoticism, fertility and unknowability with which European accounts of the colonial world were often saturated.

It was through the discursive elements of imperialist strategies that impe-rialist practices were justified and legitimated to the colonizers as well as to the colonized. However, the strategies were far from being all one way, and wherever imperialism went, resistance followed.

Anti-colonial strategies and the end of the formal empires

Western imperialism was thus produced through a diversity of strategies, some military, some economic, some discursive. Appropriately enough, it was opposed and challenged by a similarly wide range of strategies and tactics on the part of colonized peoples. Since they were undertaken, by definition, by groups and individuals who occupied subordinate positions in the social hierarchy, and who did not always have any need or wish to document their activities, our knowledge of the forms of opposition to colonial rule is less detailed than that of the strategies of the rulers.

For the most part, therefore, the accounts we are left to work with tend to be told from the perspective of the colonizing power. Even where this gives due weight to the process and practices of anti-colonial resistance, the episodes referred to tend inevitably to be those which loomed largest in the minds and lives of the colonizers, such as those in which violence was involved, for example. Hence the armed Indian rebellion of 1857 (referred to as the Indian 'Mutiny' by British imperialists) gained an especially prominent place in the official history of the British empire. I do not mean to downplay the significance of such events, and armed revolutions were central to the ending of many cases of colonial rule. The fact that they dominate the history books, however, obscures other more mundane and everyday, but often potent, forms of resistance to imperial rule.

In the case of Kenya, in East Africa, Anton Gill documents the following story from 1907, retold by the daughter of the President of the Kenyan Colonists Association:

> I think it was my mother and her sister-in-law – my father's oldest sister – who were trying to take a couple of rickshaws from the centre of Nairobi to go back to their houses in Muthaiga, and my rather tiresome aunt obviously said something that upset a rickshaw boy and he let go of his handles, with the result that my aunt was thrown out of the rickshaw backwards, and of course in those days that was an absolutely unthinkable thing to do.[27]

The brutality of the subsequent punishment meted out to the 'offender', and thousands like him in similar circumstances, helps to explain the resentment in Kenya which later led to the violent uprisings of the so-called Mau-Mau rebellions. Even an otherwise sympathetic commentator such as Victor Kiernan recoils at the anti-colonial strategies of Mau-Mau:

> Because the fighters came mostly from the poorest, most illiterate strata, with scarcely any leadership of more modern outlook, the rising took on the aspect, like many earlier ones in Afro-Asia, of a religious or magical cult. 'Mau-Mau' had some grotesque features which

made it easy to denigrate the whole movement as a relapse into an abysmal past. Rebel weapons were at first not much more up to date, a few home-made guns the best of them. Nothing like a regular force emerged from the guerrilla bands.[28]

Yet, from the point of view of the participants, the Mau-Mau uprisings, and the rituals and oath-taking surrounding them seemed very different. According to one of those involved they reflected everyday aspects of Kikuyu life:

> The *Muma wa Thenge* (the he-goat oath) is a prominent feature of our social life, an integral part of the ceremonies uniting partners in marriages, in the exchange or sale of land (before the Europeans came, when land was plentiful, the sale of land was almost unknown), or in transactions involving cattle or goats. The warriors also took an oath, known as *Muma wa Aanake* (the oath of the warriors) to bind them before going on a raid. The purpose of all these oaths was to give those participating a feeling of mutual respect, unity, shared love, to strengthen their relationship, to keep away any bad feelings, and to prevent disputes.[29]

In due course, African demands for independence, supported by more or less violent campaigns of resistance, began to be met. India had already gained its independence in 1947, following a prolonged campaign of civil disobedience and unrest, which was largely, although not wholly non-violent, in line with the philosophy (drawn from Indian religious teaching) of Mahatma Gandhi. France lost her major overseas possessions in wars during the 1940s and 1950s.[30] Portugal, with the longest-standing imperial territories in Africa, held out the longest, and with great bitterness. However, by the mid 1970s, Africa had seen the back of Portugal too. With the ending of white minority rule in South Africa in 1994, five centuries of formal white domination of Africa came to an end. According to D. K. Fieldhouse: 'Nothing in the history of the modern colonial empires was more remarkable than the speed with which they disappeared. In 1939 they were at their peak: by 1981 they had practically ceased to exist.'[31]

Post-colonialism

A dilemma

The ending of formal political control is only part of the picture, however. Other commentators have remarked on the process of informal imperialism in which economic advantages continue to accrue to the metropolitan power in the absence of direct rule.[32] Here I want to focus once more on the issue of the discursive construction of colonial relations and conclude the chapter by considering the recent development of so-called 'post-colonialism'.[33]

In writing the previous section, I found that one of the difficulties with trying to identify anti-colonial strategies is that they tend to involve forcing the histories and geographies of colonized peoples into the story as told by the West. This problem has been considered in detail by Partha Chatterjee, an Indian political philosopher. Chatterjee argues that in a situation of imperial domination even discourses of resistance and nationalist dissent are caught up in the Western world view which they seek to repudiate.[34] Derek Gregory makes a similar point:

> As I understand it, Indian historiography has been dominated by two (not one) 'imperial histories'. On the one side is the modern, secular history that the British brought to the subcontinent, through which India is ushered from brigandage and feudalism into capitalist modernity under the tutelage of the Raj. On the other side is a nationalist historiography, which casts a native Indian elite in an heroic role, wresting the state apparatus from the imperialists and completing the political trajectory inaugurated by the British.[35]

Although the two versions of history are opposed to one another, argues Gregory, from a broader perspective they are both telling the same story, of the movement of India towards a modern, ordered future. Neither, therefore, has much room for alternative ways of being and becoming, which fall outside of the imperial story altogether.

Post-colonialism and geography

I referred above to the cultural practices of the Kikuyu people of Kenya. It is clear from the firsthand account of oath-taking that one of the things which made such resistance effective was a refusal to assimilate to the European world view, which regarded such activities as barbaric. It is this difficult and complicated relationship between the ways of being, doing, thinking and speaking of the West on the one hand and of the people of its former colonies on the other which is the focus of post-colonialism.

Post-colonial writers and thinkers argue, among other things, that formal decolonization is not in itself enough. They suggest that there was much more to imperialism than political and military control, and that the domination of much of the world by Europe was also a domination by European ways of thinking and of understanding that world. One post-colonial writer, the novelist Ngugi wa Thiong'o, captures this perspective in the title of his book, *De-colonizing the Mind*.[36]

Jonathan Crush suggests that there are four elements to current attempts to write geography from a post-colonial perspective:

> the aims of a post-colonial geography might be defined as: the unveiling of geographical complicity in colonial dominion over space; the character of geographical representation in colonial discourse; the de-linking of local geographical enterprise from metropolitan theory and its totalizing systems of representation; and the recovery of those hidden spaces occupied, and invested with their own meaning, by the colonial underclasses.[37]

Let us briefly unpack each of these in turn. The injunction to examine 'geographical complicity in colonial dominion over space' implies that geographers should critically consider the ways in which geographical knowledge and skills have been (and continue to be) used to implant colonialism and imperialism in practice. Exposing 'geographical representation in colonial discourse' means showing how the discourses of colonialism involved a particular understanding of geography and particular depictions of places and regions. I discussed some of the ways in which geography has supported imperialism in the first part of this chapter.

Crush's third element is the 'de-linking of local geographical enterprise from metropolitan theory and its totalizing systems of representation'. The proposal here challenges the ways in which geography itself has been subject to colonialism. Perspectives, theories and frameworks developed in the West are widely adopted by geographers working throughout the world. Post-colonialism suggests that geographical knowledge developed in different local contexts should not be based on the assumption that Western frameworks of understanding are the only, or the best, ways of describing and understanding the world. For example, Crush states that in South Africa until recently even radical geographers have tended to adopt theoretical ideas, such as world-systems theory and structural Marxism, which were developed in Europe and North America. Lately, he suggests this situation has improved with geographers at three universities in particular shaking off some of these 'colonial' theories.[38]

According to Crush, the fourth component of post-colonialism in geography is the 'recovery of those hidden spaces occupied, and invested with their own meaning, by the colonial underclasses'. What are involved here are attempts to write geography in ways which give full weight to the experience of those who suffered under colonialism and to the places in which they live and work. This might involve, for example, considering the lives, places and political strategies of Kikuyu people on their own terms, rather than from the perspective of the imperial rulers. Crush reports that in South Africa, geographers' moves away from Euro-American theoretical perspectives have been accompanied by an increased interest in, and involvement with, local people and local political struggles.[39]

It is still relatively early in the development of post-colonial ideas, especially in geography. Many of the concepts are complex and controversial, but they are of great significance for political geography because of the stress they place on the connections between power, ideas and the unequal relationships between places and the people who live in them. These characteristics are shared by some related but rather different work on the global political order, which forms the subject of the next chapter.

Notes to Chapter 4

1 Diane Elson, 'Imperialism', in Gregor McLennan, David Held and Stuart Hall, eds, *The Idea of the Modern State* (Milton Keynes, Open University Press, 1984), pp. 154–82.
2 Karl de Schweinitz Jr, *The Rise and Fall of British India: Imperialism as Inequality* (London, Methuen, 1983), p. 39.
3 Daniel J. Boorstin, *The Discoverers* (Harmondsworth, Penguin, 1986), p. 175.
4 Eric R. Wolf, *Europe and the People without History* (Berkeley, CA, University of California Press, 1982), pp. 131–57.
5 David Livingstone, *The Geographical Tradition: Episodes in the History of a Contested Enterprise* (Oxford, Blackwell, 1992), pp. 216–59.
6 David Stoddart, *On Geography and Its History* (Oxford, Blackwell, 1986), p. 29. Emphasis added.
7 *Op. cit.*, p. 33.
8 Livingstone, *Tradition*, p. 222.
9 *Op. cit.*, pp. 224–5. Livingstone is quoting from A. Austin Miller, *Climatology* (London, Methuen, 1931), p. 2.
10 T. Griffith Taylor, 'Racial Geography', in T. Griffith Taylor, ed., *Geography in the Twentieth Century* (New York, Philosophical Library, 3rd edn 1957), pp. 455, 454. (Quoted in Livingstone, *Tradition*, p. 230.)
11 Paul Carter, *The Road to Botany Bay* (London, Faber and Faber, 1987), p. 113.
12 *Op. cit.*, p. 67.
13 Felix Driver, 'Geography's Empire: Histories of Geographical Knowledge', *Environment and Planning D: Society and Space* 10 (1992), pp. 23-40; Anne Godlewska and Neil Smith, eds, *Geography and Empire* (Oxford, Blackwell, 1994).
14 Peter Taylor, 'The Error of Developmentalism in Human Geography', in Derek Gregory and Rex Walford, eds, *Horizons in Human Geography* (Basingstoke, Macmillan, 1989), pp. 303–19
15 Anthony Giddens, *The Nation-state and Violence* (Cambridge, Polity Press, 1985), pp. 161–71.
16 *Op. cit.*, pp. 167–8.
17 Peter Taylor couches his account of world-systems analysis in a way which avoids functionalist phrasing. For example, in the second edition (1989) of *Political Geography* he writes that the three-tier structure help to stabilize the system and prevents confrontation; therefore 'those at the top will always manoeuvre for the "creation" of a three-tier structure whereas those at the bot-

tom will emphasize the two tiers of "them and us"' (p. 10). However, while such 'divide and rule' strategies are undoubtedly pursued by powerful groups, it is difficult to see how they could work at the scale of the world-economy which, by definition, is marked by the fragmentation and distribution of political power among a plethora of states.

18 D. K. Fieldhouse, *The Colonial Empires: A Comparative Survey from the Eighteenth Century* (Basingstoke, Macmillan, 1981), p. 381.
19 *Op. cit.*, p. 381.
20 *Op. cit.*, pp. 382–6.
21 *Op. cit.*, p. 382.
22 V. I. Lenin, *Imperialism: The Highest Form of Capitalism* (Moscow, Foreign Languages Publishing House, 1915).
23 J. A. Mangan, ed., *Making Imperial Mentalities: Socialisation and British Imperialism* (Manchester, Manchester University Press, 1990).
24 Stuart Hall, 'The West and the Rest: Discourse and Power', in Stuart Hall and Bram Gieben, eds, *Formations of Modernity* (Cambridge, Polity Press, 1992), pp. 275–331.
25 Anton Gill, *Ruling Passions: Sex, Race and Empire* (London, BBC Books, 1995).
26 Philip Gibbs, *The Romance of Empire* (London, Hutchinson, undated), p. 5.
27 Quoted in Gill, *Ruling Passions*, p. 114.
28 V. G. Kiernan, *European Empires from Conquest to Collapse, 1815–1960* (Leicester, Leicester University Press, 1982), p. 221.
29 Josiah Mwangi Kariuki, 'The "Mau-Mau" Oath', in Elie Kedourie, ed., *Nationalism in Asia and Africa* (London, Weidenfeld and Nicolson, 1970), p. 469.
30 Anthony Clayton, *The Wars of French Decolonization* (London, Longman, 1994).
31 Fieldhouse, *Colonial Empires*, p. 395.
32 See, for example, A. G. Frank, *Capitalism and Underdevelopment in Latin America* (New York, Monthly Review Press, 1969).
33 Stuart Corbridge, 'Marxisms, Modernities, and Moralities: Development Praxis and the Claims of Distant Strangers', *Environment and Planning D: Society and Space* 11 (1993), pp. 449–72; Derek Gregory, *Geographical Imaginations* (Oxford, Blackwell, 1994), pp. 133–205; Timothy Mitchell, *Colonizing Egypt* (Cambridge, Cambridge University Press, 1988); Edward Said, *Orientalism: Western Conceptions of the Orient* (Harmondsworth, Penguin, 1978) and *Culture and Imperialism* (London, Chatto and Windus, 1993); Gayatri Chakravorty Spivak, 'Can the Subaltern Speak?', in Cary Nelson and Lawrence Grossberg, eds, *Marxism and the Interpretation of Culture* (Chicago, University of Illinois Press, 1988), pp. 271–313; Robert Young, *White Mythologies: Writing History and the West* (London, Routledge, 1990).
34 Partha Chatterjee, *Nationalist Thought and the Colonial World: A Derivative Discourse?* (London, Zed Books, 1986).
35 Gregory, *Imaginations*, p. 183.
36 Ngugi wa Thiong'o, *Decolonizing the Mind* (London, James Currey, 1986).
37 Jonathan Crush, 'Post-colonialism, De-colonization, and Geography', in Anne Godlewska and Neil Smith, eds, *Geography and Empire* (Oxford, Blackwell, 1994), pp. 336–7.
38 *Op. cit.*, p. 340.
39 *Op. cit.*, p. 341.

|5|

Interpreting the new world (dis)order

Overview

This chapter focuses on another set of power relations at the international scale. 'Geopolitics' is a term commonly used in political geography to designate relations of rivalry between states. In what follows I will focus on the geopolitical conflict which dominated the international scene for 40 years after the Second World War, namely the unstable and dangerous balance of power between the United States and the former Union of Soviet Socialist Republics. To start with, I outline the emergence of a world order based on superpower rivalry after 1945 and examine how the so-called 'cold war' developed until the mid-1980s. I then consider the crucial impact of one man, Mikhail Gorbachev, Soviet leader from 1985, and his policies of reform which led eventually to the end of the cold war and to the breakup of the Soviet Union itself, before outlining some of the new (and not so new) geopolitical problems facing the world in the 1990s.

The basis on which international politics should be interpreted has been a lively area for debate within political geography, and in subsequent parts of the chapter I examine the issues at stake. After considering the ways in which concepts of order and disorder are used in discussing geopolitical change, I turn to some of the main theoretical approaches involved. Traditional studies of international relations have been based mainly on 'realism', and much thinking with the subject of political geography has been implicitly or explicitly realist. However, more recent work has sought to challenge this perspective, in part because of the ways in which it conceives of state power and ignores the role of non-state actors in the international system, and in part because of the deep connections between realist ideas and the practical political strategies of governments. In contrast to realism, new approaches focus also on the role of the increasingly integrated world economy on the one hand, and on the significance of discourse to the practice of geopolitics on the other.

East and West 1945–85

The legacy of war

Whichever approach is adopted to interpret the changing pattern of international relations, it is not possible to ignore the impact of the Second World War. The impact of war was paradoxical. On the one hand it had involved co-operation between states which soon came to see one another as enemies, and conflict between subsequent allies. Thus the Soviet Union fought with the United States, Britain and France against Germany, Italy and Japan. Yet within a decade, the United States and the Soviet Union were seen as global superpowers with mutually opposed interests, while Italy, Japan and the larger part of Germany became fully part of the West, economically, if not militarily. On the other hand, in spite of this rapid abandonment of wartime alliances, the war and its aftermath conditioned to a great extent the shape of the post-war world: its institutions, its forms of international relations and its economic paths.

Relations between the governments of the United States and the Soviet Union had been at best cool and at worst openly hostile before 1939. Co-operation during the war was effectively an instrumental matter carried on by both sides in the face of a threat to their respective interests, but built on opportunism rather than deep respect. The military defeat of the German and Japanese Axis saw large parts of the world occupied by the armed forces of the victorious Allies. Throughout the war, however, each of the Allies had fought in particular areas. The United States, being dominant in the Pacific, took control of Japan, for example, while Europe and German-held territory was divided between the Soviet Union, whose forces had advanced from the east, and Britain, France and the United States, moving in from the west. By 1945 war-ravaged continental Europe was being administered by the four powers, each with responsibility for particular territorial areas.

The government of the United States was particularly concerned to establish free trade, based on the capitalist system of production, as widely as possible. Immediately after of the war the American government thought that the best way to set up an international system along free-trade lines was to maintain a degree of co-operation and openness in dealings with the Soviet Union. To that end, towards the end of the war, the US government had been instrumental in setting up the key post-war international financial institutions. These organizations were established following a conference at Bretton Woods on the north-east coast of the United States. The conference, between Britain, Canada and the United States, founded the International Monetary Fund (IMF) and the International Bank for Reconstruction and Development (the World Bank). Together they were known as the 'Bretton Woods' system. The purpose was to provide a means of stabilizing the international economy on the basis of monetary stability, so that, in theory, free trade could continue without being disrupted by the political manipulation of currencies. The IMF provided loans to help governments deal with balance of payments difficulties, so that they would not need to devalue their currency relative to those of their economic competitors. The World Bank was initially focused much more on helping to finance European reconstruction after the war. Again, the aim was to get the European economies going again as quickly as possible on the basis of the free-trade system which the US government supported. In due course, however, the World Bank became the principal multilateral agency through which funding for development in the so-called Third World was allocated by Western governments.

One legacy of the war was, therefore, the territorial spheres of influence of the emerging superpowers of the US and the USSR. In addition, however, the war had a number of other impacts. First, it had involved the first use of atomic weapons in armed conflict, when bombs were dropped on the Japanese cities of Hiroshima and Nagasaki. Conflicts based on the threat or implied threat of the use of nuclear weapons were to play a major part in the subsequent development of superpower relations. Second, it set the scene in due course for the founding of the European Community (now the European Union), which, it was argued, would unite the economic interests of France and what was then West Germany, and thus prevent the possibility of future wars between them. Third, it saw the establishment of the United Nations Organization (UNO). Following a meeting in 1944 between representatives of the governments of the United States, China and Britain, a draft charter was drawn up, and this led in due course to the inauguration of the UNO on the 26 June 1945. It included a Security Council, a General Assembly and a Secretariat, as well as a number of more specialist agencies, which grew in number in subsequent years to include the World Health Organization (WHO), the Food and Agriculture Organization (FAO) and the United Nations Children's Fund (UNICEF).

Much of the decision-making power of the UNO was vested in the Security Council, on which the United States, France, Britain, China and the USSR sat as permanent members. Each of these governments had a right of veto within the Council, leading to many occasions on which proposals for action were stalled because they were thought by one or other of the permanent members to run counter to the interests of their government.

The rise of the post-war order

The tentative co-operation which had existed between East and West at the end of the war did not last long. The Bretton Woods arrangements were administered by the participating governments, with voting rights being allocated according to the financial input of each country. This meant that the United States, as the richest donor, had by far the largest single say in the running of the institutions. The Soviet government correctly regarded the Bretton Woods system, and the policy of open trade, as attempts by the United States to open up as much of the world as possible to capitalism in general and to its industries in particular. By the same token it was clearly intended, as the American government saw it, to 'contain the spread of communism' from the Soviet Union to other countries and especially to Western Europe. In response the Soviet Union set out to secure what it saw as its sphere of influence and to surround itself with friendly satellite states which would act as a buffer between the Soviet territory and the rest of the world. In 1946, Winston Churchill used the term 'Iron Curtain' to refer to the division which separated Europe into a Western zone allied to the United States and an Eastern zone allied to the USSR.

In 1947, the aim of preventing Europe from 'falling' to communism was made explicit when a special scheme was proposed by the then American Secretary of State, George Marshall. This plan, known as the 'Marshall Plan', pumped $17 billion into Western Europe to finance its reconstruction and to tie it, economically and politically, to the capitalist system. In the same year, the USSR consolidated its hold over Eastern Europe, sponsoring and supporting communist-led governments. This organization of political space was gradually deepened over the coming years.

1948   The formation of the Organization for European Economic Co-operation (OEEC – later incorporated into the Organization for Economic Co-operation and Development).

Soviet blockade of Berlin, which had been divided between the four powers of France, Britain, and the United States and the USSR. The Western allies organized an airlift to keep the Western sector of the city supplied, and defeated the blockade.

1949   Formation of the North Atlantic Treaty Organization by twelve Western governments.

Formation of the Council for Mutual Economic Assistance – the organization linking the economies of Eastern Europe to the Soviet central planning system.

1951   Establishment of the European Coal and Steel Community – forerunner to the European Economic Community.

1955   Formation of the Warsaw Pact defence alliance for the Soviet bloc.

The first cold war and the geopolitics
of containment

The development of these institutional and political frameworks seemed to be sedimenting a static ordering of the world around the two superpowers. This first cold war was dominated by mutual and growing distrust between the USSR and the United States. From the perspective of the USSR, the United States was an expansionist, imperial power against which it had to defend itself militarily and ideologically. From the Soviet perspective the Warsaw Pact countries were an essential buffer between the Soviet mother-land and a hostile outside world. While the United States had only two international borders, both with friendly countries, the USSR, in the middle of the Eurasian land mass, was surrounded by what it perceived as a ring of potentially hostile states.

From the American perspective, the buffer zone around the Soviet Union was the beginning of Soviet expansion and an expression of Soviet goals for world domination. This perception was lent some credence by Soviet sup-port for nationalist movements in Africa, Asia and Latin America. American concern about the supposed expansionist inclinations of the Soviet govern-ment led to a policy of 'containment'. The doctrine of containment was expressed by US State Department official George Kennan: 'it is clear that the main element of any United States policy toward the Soviet Union must be that of a long-term, patient but firm and vigilant containment of Russian expansionist tendencies.'[1]

This led to the establishment of a ring of US-backed alliances around the territory of the Soviet Union, serviced by a network of US military bases. The alliances included, in addition to NATO, SEATO (South East Asia Treaty Organization), CENTO (The Central Treaty Organization) and ANZUS (the Australia–New Zealand–US pact). This policy of containment, of course, tended to exacerbate Soviet fears that the United States was hos-tile to the USSR.

With the death of Stalin and the coming to power of Khrushchev in 1955, US–Soviet relations entered a new phase marked by considerable instability. Initially, things began well, with Khrushchev's declaration of a policy of mutual co-existence in which, he suggested, each superpower should respect the right of the other to hegemony within its sphere of influence. Over the following 15 years, the cold war went through a series of thaws and re-freezings. A series of diplomatic incidents leading to periods of heightened tension were interspersed with periods of accommodation, occasionally leading to meetings between Soviet and American leaders. Probably the most significant and potentially dangerous incident was the Cuban missile crisis of 1962 in which America blockaded Cuba to prevent the Soviet gov-ernment installing nuclear missiles on the island. Soon afterwards, however, there was a degree of rapprochement with the Partial Test Ban Treaty in

1963 and the decision to set up the so-called hot-line between Washington and Moscow. This accommodation did not last long, however, as the United States intervened militarily in Vietnam. The US government tried to justify the Vietnam war on the grounds of the containment policy, arguing that the communist government of North Vietnam was part of the expansion of world communism and had to be contained to prevent the spread of communism throughout south-east Asia. In the event, the United States, then, as now, the most powerful military state in the world, was unable to beat the impoverished and relatively ill-equipped North Vietnamese army.

Détente

During the 1970s, relations between the two superpowers improved considerably. Richard Nixon inaugurated his Presidency of the United States with a proposal to move from confrontation to negotiation. Throughout the 1970s, there were a series of summits between the Soviet leader Leonid Brezhnev and successive American Presidents. In addition there was a growth in cultural and economic links between East and West and a series of negotiations leading to an agreement to limit strategic nuclear arsenals (SALT – Strategic Arms Limitation Talks).

This period was known as '*détente*' and involved more than just the superpowers. In Germany, for example, a number of moves were made to develop closer links between the Federal Republic in the West and the Democratic Republic on the other side of the Iron Curtain in the East. Furthermore in the South there were signs that the alignments between many developing countries and one or other superpower were breaking down. It could no longer be taken for granted that the rest of the world would line up neatly on one side of the East–West divide or the other. Nevertheless, superpower conflict in the South did not cease overnight, and at the end of 1979 the Soviet Union invaded Afghanistan, an event which marked the beginning of a new cold war.

The new cold war

The United States government and its President, Jimmy Carter, reacted dramatically. Contacts between the two countries were scaled down, the SALT II treaty was withdrawn from the American Senate and the United States led what it hoped would be a world-wide boycott of the 1980 Moscow Olympic Games. In terms which became known as the 'Carter Doctrine' the

President stated that the American government would regard any attempt by the Soviet Union to gain control of the Persian Gulf area as a threat to the vital interests of the United States. 'Such an assault', said Carter, 'will be repelled by any means necessary, including military force.'[2]

This inaugurated a new phase of hostility, which was, if anything, intensified when Ronald Reagan assumed the US Presidency in 1981. Reagan began by embarking on a programme of expanding military spending. Coming from America's radical right wing, he argued that the Soviet Union was a threat to world peace, and took a markedly more confrontational attitude to superpower relations than his predecessors. In Europe, heightened tension involved the deployment of Soviet and American nuclear-armed cruise missiles on the territories of their respective allies, moves which were greeted by significant increases in membership and activity of grassroots peace and nuclear disarmament campaigns. The sour relations between East and West dominated many international issues during the 1980s. In Central America, for example, the left-wing Sandinista government of Nicaragua faced an ongoing campaign by US-backed 'Contra' guerrilla forces attempting to destabilize and overthrow what the American government regarded as a communist threat in 'its own backyard'.

However, although there was as yet no return to the cordiality of *détente*, 1984 marked a noticeable softening in Reagan's rhetoric towards the Soviet Union. Although it was not obvious at the time, when Mikhail Gorbachev became General Secretary of the Communist Party of the Soviet Union in March 1985, the stage was set for some dramatic transformations in the geopolitical world order.

East and West 1985–95

The Gorbachev revolution

When Gorbachev came to power, he was aware that years of neo-Stalinist bureaucratic planning and the prioritization of military security had undermined the productiveness of the Soviet economy and had led to the stagnation of Soviet society. He initiated a series of reforms which led eventually to a wholesale revolution in the political and economic organization of the Soviet bloc. The strategy he adopted was threefold.

First, *perestroika*, or restructuring, referred mainly to the reorganization of economic life. The role of central planning was significantly curtailed, subsidies to industry were cut back, a small co-operative sector was encouraged and joint ventures with foreign investors were allowed. Eventually the changes were extended to include the private ownership of the means of production, thus opening up the Soviet economy, in principle, to the expansion of genuinely capitalist relations of production. Second, *glasnost*, or openness, allowed the Soviet people increased freedom of expression, with censorship abolished in many areas, and open debate and criticism allowed for the first time. Third, *demokratizatsiya* (democratization) opened elections to more than one candidate and removed the rule that candidates should be members of the Communist Party.

These three sets of changes generated much opposition, particularly from those in the Communist Party who wanted to retain the existing system. It was clear that, at least at first, Gorbachev wanted to hold the territory of the Soviet Union together, and thought that he could do so. In due course, however, it became clear that the reform processes which he had begun were going to have consequences well beyond his expectations.

The fall of the wall

In 1961 the East German government, concerned at the number of people fleeing to West Berlin, built a wall between the Eastern and Western sectors of the city. The Berlin wall symbolized the division of Europe in a dramatic fashion, and the breaching of it in 1989 amidst popular demonstrations was an equally powerful symbol of the geopolitical changes under way.

In the late 1980s, Gorbachev hinted that the Soviet government would no longer use military force to maintain communist governments in its East European satellite states. By 1989 this commitment was explicit, and the way was clear for reforming and popular movements to come to power. In the period since, all the former state socialist countries in Europe have developed representative forms of electoral government. At the same time their economies have been subject to significant, and significantly different, transformations, with the end of the system of central planning.

1989 will undoubtedly go down as one of the most dramatic years in modern European history. It began with communist governments in power throughout Eastern Europe. In June, the Solidarity movement led by Lech Walesa won free elections in Poland. Through the autumn of 1989 a succession of popular revolts fatally weakened or toppled communist governments in Hungary, East Germany, Bulgaria, Czechoslovakia and Romania. In some cases, such as Czechoslovakia, the transition was relatively peaceful. In Romania, by contrast, a violent coup included the execution of the communist leader, Nicolae Ceausescu.

One important effect of the revolutions of 1989 has been to expose marked underlying cultural and economic differences between the various countries of central and eastern Europe. Czechoslovakia has now split into two separate states: Slovakia and the Czech Republic. Of these, the Czech

Republic is considerably more advanced, industrially, and it may in due course join the European Union with Hungary and Poland. Other countries such as Romania, Slovakia and Bulgaria have been less well placed to capitalize economically on their freedom from Soviet influence.

The other impact of Gorbachev's reform programme was the breakup of the Soviet Union itself.[3] To date this has effectively meant disintegration into the fifteen national republics which made up the USSR, although in some cases, there are nationalist minorities within those republics which are pressing for their own independence. Of the former Soviet republics, the Baltic states of Lithuania, Latvia and Estonia were able to press their claims for independence most effectively. Having recently been sovereign states themselves prior to the Second World War, and being among the most economically developed parts of the Soviet Union, Gorbachev found it difficult to resist their moves for autonomy for long. In other parts of the Soviet Union, especially in Soviet central Asia, there were fewer resources for state-building and fewer gains to be made from outright independence. Here the republics have remained in a loose alliance with Russia, known as the Commonwealth of Independent States.

A new world order?

The changes inaugurated by Gorbachev, the ending of communist rule in Eastern Europe and the breakup of the Soviet Union led to the final collapse of one half of the old bi-polar world order. The United States found itself, in geopolitical and military terms at least, unchallenged as the world's only superpower. A US State Department official, Francis Fukuyama declared in 1989 that history had ended: the West had won the cold war, and this represented a victory for capitalism, liberal democracy and free-market economics.[4] The argument that the end of the cold war was a victory for the West was hotly disputed, but for a brief period it did seem as if the ending of superpower hostility might lead to a more stable and peaceful world in which international politics would be governed by negotiation and agreement, rather than the use, or the threatened use, of military force. In consequence, there was much talk during 1990, at least by the United States government, of the prospects for a new world order. The argument was that now that the United Nations was freed from the ideological deadlock, which had made it so difficult for it to act in the past, it would be able to assume its true purpose of securing the peaceful mediation of international conflicts. This role would be underwritten by the United States, which seemed poised to appoint itself as a global police force.

Not all countries were happy with the idea (at least in its US-dominated version), but in any case it did not survive for long. The Iraqi invasion of Kuwait in 1990 led to the imposition of international sanctions against Iraq by the United Nations. However, this had no effect, and it was only when the United States persuaded the UN to agree to the use of military force and led a counter-invasion that Iraq was forced to withdraw. Moreover, it has become increasingly clear that the United States, while concerned as ever to protect its own perceived vital interests, is increasingly unwilling to provide the resources to underwrite global military security in any general way.

Multi-polar geopolitical disorder?

An increasingly popular, and probably more realistic, assessment of international relations in the contemporary world is that they are characterized not by an emerging order, but by disorder. The transition from state socialism was marked in Yugoslavia, as it then was, by the development of the first major war on the continent of Europe for 50 years. In the territory of the former Soviet Union, other nationalist wars have marred progress towards greater political freedom. In other parts of the world, such as southern Africa, armed conflicts have been ended in favour of negotiation or the peaceful transfer of power. But whether disputes are settled militarily or otherwise, there is certainly scant evidence for the development of the kind of geopolitical stability which would merit the label 'new world order'.

In some respects this is unsurprising. For 40 years, the East–West balance dominated global politics and structured thinking about international issues. With its end, a neat (in some senses) bi-polar world has been replaced by a much messier multi-polar one. This need not necessarily involve widespread military conflict, but it certainly raises a wide variety of conflicts of interest which are either new, or have been suppressed during the cold war.

In a multi-polar world there are many actors, all of whom have some ability to influence events. Thus power is more widely distributed, but conflicts and rivalries are liable to proliferate. In terms of states, the dominance of the US and the USSR is being replaced by a plethora of other actors. The United States remains important, of course, as does Russia, but they are increasingly being joined by a diverse band of other states, and groups of states. These include Japan, the fast-growing economies of south-east Asia, Germany and the European Union and China. In addition, non-state actors are, as we shall see below, of increasing importance, including multinational corporations and intergovernmental organizations.

Contemporary issues in international politics

The proliferation of significant actors on the international political stage is matched by a growth in the range of issues and problems facing them. Some of these issues are hardly new, of course, but with the ending of the dominance of the US–Soviet conflict, they have moved up the agenda. In their account of the new multi-polar world, Charles Kegley and Gregory Raymond identify the following key issues:

1 the increasing destructiveness, accuracy, and proliferation of modern weaponry and, despite the end of the cold war, intensified thirst for security through military preparedness;
2 the continuing internationalization of national economies;
3 the passing of political power from governments to private transnational actors such as multinational corporations;
4 the widening gap between the world's rich and poor, exacerbated by exponential population growth among those impoverished countries lacking an adequate technological infrastructure;
5 the unabated deterioration of the global ecosystem; and
6 the resurgence of hypernationalism and outbreak of civil wars bred by ethnic conflicts.[5]

To these we might also add:

7 the growth of international organized crime, particularly in relation to trade in illegal drugs and weapons;
8 the growth of religious movements, including some prepared to use violence to pursue their goals;
9 the globalization of media and communications technology.

These, then, are some of the most important political problems at the international scale at the end of the twentieth century. How they are dealt with will depend on political conflicts and alliances which are still emerging. In the second half of this chapter I want to consider how such international political processes might be interpreted.

Order and disorder in international affairs

So far I have used the terms 'order' and 'disorder' to contrast the geopolitical arrangements established after the war with those of today. In the 1960s and 1970s the world was often represented as ordered, because different sides seemed to know where they stood in relation to the others, even if it was an order based on the threat of mutual nuclear annihilation. In the

1990s the world feels to many people and governments to be much more chaotic politically, as a wide range of different interests, states, social movements and individuals attempt to exert influence and to pursue strategies in the international arena. Of course, in many cases these different actors and interests are not at all new. During the various phases of the cold war, however, they tended to be subordinated to a broader conflict between East and West. With the end of that stand-off, at least in anything like its old form, issues such as nationalism, international crime, the environment and religious revivals have come to the fore.

However, the terms 'order' and 'disorder' are ambiguous. As descriptions of the world they are not neutral value-free terms. As well as being terms used by social researchers and political thinkers to talk about the world, they enter into the discourse of practical politics too. Very often, for example, 'order' is assumed to be a 'good thing' in itself, without much consideration being given to the relations of inequality or oppression which might be built into the structure of that 'order'. An ordered world makes people and especially governments feel secure, because things seem more predictable and thus easier to deal with. By contrast, throughout human history, the notion of disorder has gone hand in hand with feelings of threat, and fears about the future.

Political thinkers and writers therefore need to take great care with how these terms are used. Much writing on international relations and geopolitics has involved attempting to identify the nature of 'order' in international politics; a variety of models and theories of the 'international world order' have been produced over the years. However, an understandable desire for an ordered world can lead to over-emphasizing the completeness and coherence of political order. For example, governments in the West often argued that the so-called 'balance of power' between NATO and the Warsaw Pact until 1989 underpinned an ordered world in which Europe had remained peaceful for 40 years. On the other hand, others have suggested that that 'peace' was actually highly unstable, based on a delicate nuclear balance which could easily have tripped over into nuclear war, and which, on more than one occasion, very nearly did. At the same time, the desire to believe in an ordered world led wealthy governments, and the political theorists and writers whose ideas were the basis of their thinking, to neglect the majority of the world's places and peoples who were leading lives of extreme poverty. Throughout much of what was called the 'Third World', wars were fought using weapons and military expertise imported from the United States and the USSR, and expressing conflicts of interests between the superpowers. For those caught up in such wars, in Korea, in Central America, in south-east Asia and in parts of sub-Saharan Africa, the world must have seemed far from ordered.

There are also problems for those who stress a world ordered on the basis of increasing economic integration. For sure, the 40 years after the end of the Second World War saw global economic expansion on a huge scale, with production and finance being more and more internationalized and world markets trading an ever-growing quantity of commodities. Yet this integration is not complete, nor does it imply an increased orderliness in human affairs. For millions of people throughout the South, economic activity is rural, not urban, and remains based on a combination of subsistence production, part-time and temporary wage labour in agricultural or small-scale industrial production, and limited trading in local, rather than global, markets. World events, and global-scale processes certainly have an impact on the conditions in which most people live, but they can hardly be described as harbingers of order. Strangely, Marxian writers have often been among the keenest to stress the high level of integration of the world capitalist system, although Marx himself frequently pointed to the disruption and turmoil caused through the implantation of capitalist relations of production into new areas.

The other major difficulty with the notion of order is that it can be a 'naturalizing discourse'. We have encountered this idea in previous chapters, but a quick recap maybe useful here. A 'naturalizing discourse' is a discourse which works by making a certain set of *social* arrangements, which developed through historical social and political action, seem *natural* and therefore (a) desirable; (b) inevitable; and (c) impossible to change through human action. The natural world is assumed to be ordered, and thus order in political life, rather than disorder, seems to be natural and normal. The effect of this may be to freeze an existing pattern of social arrangements, even though they are highly unequal, or oppressive to certain groups in society, and at the same time to imply that nothing can be done, or that the 'system' will develop or evolve ('naturally') to deal with the problem.

The concept of disorder is often used in similar ways. During times of anxiety or political tension or conflict, the world may be represented as 'disordered' to justify the imposition of more orderly arrangements. A fear of disorder can be detected behind many political conflicts in the modern world, including those organized around nationalism, the environment and international crime. Again the same ideas are at work, in reverse, with disorder being presented as a departure from stable normality.

The concepts of order and disorder are therefore not simple descriptive terms, used by social scientists to depict the world in an objective fashion. On the contrary, their use is a part of the world that social scientists want to depict. Writings *about* international politics and geopolitical relations become *part* of international politics, with politicians using the ideas of order and disorder in their own political strategies. This important point is not limited to international relations, of course. Increasing numbers of social researchers argue that social science does not merely reflect the social world, but also helps to constitute it, to make it how it is.

However, the study of international relations does show up particularly well this tendency for academic political thinking and research to become entangled in practical political processes. The remainder of the chapter will consider this in more detail in discussing both traditional and critical approaches to interpreting global political change.

Interpreting change: traditional approaches

In dividing the various approaches to interpretation into a traditional group and a critical group I neither mean to imply that traditional approaches have been completely superseded, nor that areas of debate and dialogue between them do not exist. Although these approaches have strong echoes in work in political geography, the distinctions between them are drawn most starkly and explicitly in the neighbouring discipline of international relations.

Idealism

The discipline of international relations grew up after, and partly in response to, the First World War. The extent of death and destruction during the war, and the widespread perception that much of it had been largely pointless even in military terms, led to concerns that everything should be done to prevent a reoccurrence. The hope of early international relations scholars was that they might contribute to a better world by promoting the peaceful resolution of international conflicts and by developing new international institutions and legal processes which would allow war in the future to be avoided. From the start, therefore, academic concerns were entangled in practical diplomatic and political matters. The desire to find ways of avoiding war also prompted the establishment of the League of Nations in 1920. The 'idealists' in the international relations field were so-called because they sought to change the world for the better, rather than trying to describe and explain it 'as it was'. The failure of the League of Nations to prevent the Second World War was paralleled by a challenge to 'idealism' from the other main traditionalist position within academic studies of international politics, 'realism'.

Realism

REALISM AND INTERNATIONAL RELATIONS

'Realism' has dominated the Western study of international politics for much of the post-war period. The use of the term is confusing for those familiar with the literature of contemporary human geography, where it is used to refer to the 'critical' realism of Roy Bhaskar and Andrew Sayer. In the international relations literature 'realism' has a very different meaning. According to the realists, the idealists' attempts to influence international relations in the directions of peace were flawed because they were not based on an understanding of the actual nature of state power and motivation. What was required, the realists argued, was an 'objective' account of the relative power of different states. However, in practice, realism was by no means the 'objective science' which it claimed to be, as we shall see.

According to realism, national states are the building blocks of international politics. All states are assumed to be similar in their motivations, namely that they are concerned first and foremost with their own survival and interests. States are viewed as individual autonomous 'agents' which interact with other agents (states) in the international arena. An implicit assumption is that states are unified and that their governments speak and act in the 'national interest'. The notion of 'national interest' is itself an interesting one. It is based on the assumptions that (1) there is a unified national interest within each state, that (2) the government of the state can identify that interest, and that (3) the government has the will and the ability to act in accordance with that interest. These three assumptions are widely held (or at least propagated) not just by realist international relations scholars, but also by many governments and their advisors. They also form the 'common-sense' view that most people have of international politics, and one which is commonly expressed in the media. For example, when reporting on international affairs, British television and radio journalists will often say 'the British argued that. . .' or 'we are being threatened by . . .', when they are actually referring to the activities and concerns of the British *government* rather than British people in general. The assumptions on which the notion of 'national interests' is based are all debatable. In practice all states contain a variety of conflicting interests and it is by no means self-evident that there is usually, or indeed ever, an overarching 'national interest' which should eclipse societies' divisions. Even if there were such an interest, it is far from clear that governments have the capacity to identify it, and they certainly do not always have the desire or the capacity to pursue it. A more plausible account of the process is that governments identify a variety of policies which *they* wish to pursue and a variety of strategies through which to pursue them. These are then *presented* as being 'in the national interest' even though they may be principally in the interests of particular social groups, or of the government itself. Moreover, different interests, policies and strategies may conflict with one another, even though a government may want to appear to be committed to them all.

Realism thus focused on a particular set of issues within international politics. These included the power politics developed by each state in pursuit of its interests, including war and the preparation for war; power politics 'simply means the stronger state imposing its will on the weaker'.[6] The theory is state-centred, so that it downplays the role of other international actors, such as firms, and intergovernmental organizations. At best these tend to be seen as instruments of state power, rather than actors in their own right. While all states claim the right to monopolize the means of violence, the realist perspective accepts that this monopoly is legitimate, and that this provides one of the key means through which states pursue their interests. The issue of state security is emphasized above other issues, such as economic motives, or relations of exploitation. Because different states have different resources at their disposal, there is a power hierarchy in international politics, but little inherent political order. Successful pursuit of state interests is thus thought to depend in part on assessments of the motives, reasoning and resources of other states, which has led some scholars to use various forms of game theory in attempts to model state behaviour and the outcomes of international conflicts.

THE CONTRIBUTION OF POLITICAL GEOGRAPHY

Traditional political geography has not put much stress on developing explicit theoretical frameworks. However, many of its implicit assumptions have something in common with the realist approach in international relations research. Within political geography, the study of international relations has usually been labelled 'geopolitics'. Peter Taylor argues that 'geopolitics has generally been part of the realist traditions of international relations.'[7] A good example is the work of Sir Halford Mackinder. I have already mentioned the role of Mackinder's role in the emergence of political geography. Taylor locates his work squarely in the 'power politics' tradition. Mackinder's 'heartland theory' was first developed in 1904 (although the term 'heartland' did not actually appear until 1919) and it remained influential in political thinking throughout most of the twentieth century, even if that influence was not always explicitly acknowledged.[8]

According to Mackinder, the great sweep of history since the Middle Ages could be interpreted in terms of the shifting balance between land-based and sea-based powers. Prior to Columbus, he argued, the initiative and power lay with those best able to command the land, and that meant the ability to travel about it quickly and easily. It is in this which Mackinder locates the source of the power of the horsepeople of the Eurasian steppe. With the opening up of the western oceans to European sea travel, the balance of power shifted towards the Atlantic where it remained for some three or four hundred years. With the development of the railways in the nineteenth century, Mackinder thought that the initiative would switch back once more to those countries occupying more continental positions. Writing during the so-called age of imperialism, Mackinder was particularly concerned about the position of Britain relative to Germany and Russia, both of whom were land-based rather than sea-based powers.

Mackinder proposed that the Eurasian land mass (which he called 'world island') was once again becoming the 'geographical pivot of history'. He coined an aphorism to express his ideas: 'He who rules Eastern Europe commands the heartland. He who rules the heartland commands World Island. He who rules World Island commands the World.' In the first quarter of the twentieth century, Mackinder used his theory to justify (among other things) Britain's maintenance of a strong overseas empire. More recently, however, the same thinking, identifying control over the Eurasian land mass as a source of potential world domination, has clearly informed such US policies as the containment approach to the Soviet Union mentioned above.

REALISM AND THE COLD WAR

From the realist perspective, the changes in US–Soviet relations during the last 40 years were the product of the shifting power balance between the two sides. Realism provided the framework for US government thinking in the sense that it starts from the assumption that the world is made up of potentially hostile states all of whom will pursue their own 'national interests' by any means up to and including the use of military force. The starting point for realist foreign policy is that the other side is assumed to be an ideological and military threat unless proven otherwise. Realism was thus less an *analysis* of the cold war than an integral part of the political *practices* of it. The ending of the cold war is interpreted in realism as an expression of the geopolitical weakness of the Soviet Union, and corresponding greater power of the United States (hence Francis Fukuyama's assertion that the West had 'won' the cold war).

Interpreting change: critical approaches

The debate between idealism and realism is often referred to as the first great debate in the study of international relations. Since then, the dominance of realism, especially in the United States, has been challenged by a variety of other perspectives. Despite the differences between them, I have grouped them together under the broad heading of 'critical approaches', because whereas realism takes the present world political map pretty much at face value, and accepts the legitimacy of state sovereignty and focuses more or less exclusively on relations between states, the other approaches call into question the present division of the world and the distribution of power within it. As we shall see, however, they do this in very different ways.

Geopolitical economy

The first group of approaches focus on economic relations of domination, dependence and interdependence in the international arena. They question, at least implicitly, a number of the assumptions of the realist perspective. First, they question the notion of the 'national interest' by arguing that international relations do not consist in pursuing 'national interests'. Rather, they suggest that what is presented as the national interest is in fact usually a *sectional* interest which is being defined as 'national' for political purposes. Second, they suggest that states are not independent actors on the international stage, interacting in an unsystematic way through threats, war, and defence. By contrast they argue that there is a degree of systematization or structuring in world affairs, which reflects more than just a contingent pattern of diplomatic alliances and conflicts, but which is organized around rather more enduring relations of shared economic interests and intersecting networks of economic processes. Third, they challenge the 'state-centric' view of realism, by 'unpacking' the economic and political make-up of states and emphasizing that other international actors, such as firms, are also important.

The roles played by non-state actors in international politics are emphasized by the work of Susan Strange. In particular, she focuses on firms which operate across international frontiers – multinational or transnational corporations (MNCs and TNCs). Such firms are actors in international politics for a number of reasons. First, their economic interests are not wholly tied to the economy of any one particular country. As capitalist economies become more open and globalized, it becomes more and more difficult to talk about 'British' firms or 'American' firms or 'German' firms. Instead there are firms with interests and operations in Britain, America or Germany, but which may have headquarters somewhere else, and, in the case of public companies, have shares which can be purchased in the stock market by people and institutions of any nationality. MNCs act (in the last instance) in the interests of their shareholders, and they owe no particular loyalty to places or national economies. They can respond in a variety of ways to changes in governments' policies which affect them. For example, they can relocate their production according to variations in wage rates between different places. They can make transfers of goods and services across international frontiers *within* the company at prices which are set so as to inflate profits artificially in one country (where profit taxes may be low) and to run up losses in another country (where there might be tax relief for loss-making businesses). Although they do not have territory in the same sense as a state, MNCs do thus have a degree of independence from governments which see themselves otherwise as sovereign.

Second, MNCs are political actors because of the huge resources which they control. According to Daniel Papp, for example, in 1988, forty-one of the world's hundred largest economic units were MNCs. Heading the list was the United States, with a total gross domestic product of $4,862 billion, followed by the then USSR, Japan and West Germany. At number twenty-one was General Motors, with an annual turnover of $121 billion, just below Mexico, but ahead of Austria, Sweden and Switzerland. Thereafter MNCs begin to dominate the list. Of the fifty units with GDPs/turnovers between $19 and $46 billion, thirty-three were firms.[9] As the number and size of MNCs grows, and their freedom to move people, goods and money around the globe increases, governments are becoming increasingly dependent on MNCs for investment, employment and taxation revenues, and MNCs are thus gaining political power from their economic resources.

Third, MNCs are political because they can and do undertake political activities. Investments by some MNCs in poor countries or in the former Soviet bloc have been undertaken on some occasions to support a particular political regime or development path. As major beneficiaries of deregulated trade and the capitalist system of production, MNCs are inclined to support pro-capitalist governments, or to withdraw investment from countries with socialist governments. Part of the interest among MNCs in making investments in Eastern Europe stems from a desire to secure a capitalist future for the region.

Susan Strange argues that the growing importance of firms in international politics stems from structural changes in the world economy, including technological developments, increasing mobility of capital and a rapid growth in the speed and volume of international communications. The last of these is significant not only because it provides the means through which firms can 'go global' but also because through telecommunications, such as satellite television, people in many places have been made much more aware of the ways of life in other areas, particularly North America. This has led to increased popular demands for access to the goods and services available in Western economies, and dissatisfaction with political and economic systems which failed to deliver them. Strange argues that this was particularly important in the fall of the communist governments of Eastern Europe and the Soviet Union. It is important to bear in mind, however, as we saw in the previous chapter, that the wealth of Western economies is in part a product of the poverty of other parts of the world, so although the demand for Western lifestyles may have grown, there is no guarantee that the increased expansion of capitalism will be able to fulfil it.

As a result of these changes, firms are increasingly pursuing political strategies of their own in the international arena. This includes, Strange suggests, both inter-firm bargaining and bargaining between firms and states. Firms have thus become more and more involved in forms of diplomacy which were traditionally viewed as the legitimate province of states. States now bargain with firms for investment and resources, but firms also bargain with each other, for example in developing international consortia to produce new products, such as large aircraft, which are beyond the scope of any one company.

John Agnew and Stuart Corbridge share Strange's emphasis on the role of economic processes and actors, and they coin the term 'geopolitical economy' to capture their concern with the relationship between geopolitical change and economic dynamics. They argue against what they regard as the traditional emphasis in geopolitics on static geographical features:

> We propose that to understand this new situation requires a new view of geopolitics, which can no longer be seen in terms of the impact of *fixed* geographical conditions (heartlands/rimlands, lifelines, chokepoints, critical strategic zones, domino effects etc.) upon the activities of the Great Powers. Rather, today *geopolitical economy* is replacing classical geopolitics as the fundamental context for the constitution of foreign policy.[10]

Their dynamic conception of international power relations neatly encapsulates some of the arguments I have been making so far and also emphasizes the relationships between material interests, political strategy and discursive representations which I outlined in Chapter 1:

> What is needed is a framework for defining geopolitics which incorporates both processes of economic and political change and the rhetorical understanding that gives a geopolitical order its appeal and acceptability. We are in a period of geopolitical disorder in which no dominant geopolitical discourse has yet emerged to match changed circumstances. The constitution of a dominant geopolitical discourse is achieved by a widely accepted division of the world into spaces of greater or lesser importance from the point of view of the contemporary Great Powers. Geopolitics, therefore, involves a process of reductive reasoning by which practitioners of statecraft constitute and read places (continents, regions, states etc.) in terms of their utility and importance to regional and global objectives and interests. States and regions are thus read geopolitically as 'clients', 'buffer zones', 'spheres of influence' or 'strategic locations'.[11]

In other words, geographic spaces and places do not have any absolute and inherent political significance, but are *made* significant through processes of discursive construction. Moreover, these meanings are likely to be contested, so that one power's 'buffer zone' may be seen by a rival power as a 'strategic location'. As Agnew and Corbridge continue:

> It is a process of representation by which political leaders (and experts) define the world and fill it with subjects, narrative and scenarios with little sense of the contradictions, ironies and pluralities, or even the histories, of particular states or regions. It is the representation of dominant powers, perhaps especially the United States after 1945, that 'script' or write the agenda for the world community at large. With the disintegration of the post-war geopolitical order we are as yet without a new script to match.[12]

Although, like most writers on the subject, Agnew and Corbridge are agnostic about the future shape of geopolitical relations they are confident that, unlike in similar situations in the past, the world's dominant power is not about to be eclipsed by the growth of a major rival. The United States, they suggest, will continue to play a central geopolitical role for four reasons:[13]

1 There is presently no single competitor which combines military and economic strength.
2 It is three times as wealthy as Japan, its nearest competitor, so that there is scope for considerable relative decline in the US economy without it losing its foremost position.
3 By comparison with the 1950s and 1960s, present-day US military commitments overseas are much smaller and more manageable.
4 Unlike in previous centuries, the great interdependence of the global economy means that economic relations are of much more significance, in comparison with military and traditional 'power political' relations. Pretenders to America's position of geopolitical pre-eminence are unlikely to pursue a political–military challenge to the US at the expense of their economic prosperity.

Following on from this last point, Agnew and Corbridge argue that economic power is no longer something which states 'possess' – something to be added in to their portfolio of resources. Rather, the globalized international economy has considerable independence from any one state or group of states. Consequently international politics is now the product of the interaction between an increasingly integrated global economy and a system of states pursuing strategic activities in relation both to other states and to economic processes which are largely beyond their control.

Agnew and Corbridge suggest that such strategies owe much to the discursive representation and 'scripting' by governments and other political actors of their geopolitical world. The implications of this insight have, in recent years, been among the most fruitful areas for research into international politics, and I will conclude this chapter with a consideration of what has become known as 'critical geopolitics'.

Critical theory and critical geopolitics

CRITICAL THEORY

The term 'critical theory' has a particular meaning in the literature of the social sciences. At first it may seem a general label, which could be applied to any theory which is 'critical' (of the working of the world, or of other theories). As it is used here, however, it refers to a particular tradition in the social sciences which questions the taken-for-granted assumptions that underpin conventional perspectives. In the international relations field, 'realism' is the main conventional perspective. Dalby cites the work of R. W. Cox to show what applying critical theory in international relations studies implies. According to Dalby, Cox draws a distinction between 'problem-solving' and 'critical thinking':

> Realism and most of international relations . . . are a matter of problem solving, or fairly narrowly defined technical endeavours. This form of thinking takes the world more or less for granted, offering suggestions for manipulation and control of events within the existing understandings and power structures. On the other hand, critical thinking penetrates further, asking questions of how the current situations came to be. . . . Instead of focusing in on small facets of a situation and subjecting them to further analytical division as 'problem-solving' theory does, critical theory opens up the investigation to explore the possibilities for change and the structural linkages between the material which is focused on and its larger context. Problem-solving theory is supportive of the status quo whereas critical thinking asks more fundamental questions of how power works and might be challenged.[14]

The realist perspective asserts that the social researcher investigating international politics is an objective, detached observer of power conflicts. Critical theory exposes this assertion as a myth.[15] The practice of international relations is not detached from students and theorists of international politics at all, but is actually the product of the theories. Put crudely, realist analyses of international politics which stressed 'power politics' helped to persuade governments to *practice* power politics. The form that international relations took was as much the product of realist theories as vice versa. Moreover, because the realist perspective was presented as objective it served to provide a supposedly scientific justification for practising politics in a particular way. The critical theory approach shows that any conceptual approach which takes the existing structures and power relations of the world for granted runs the risk of helping to perpetuate them, and thus becomes part of those power relations, and not a detached observer. While no critical theory is detached from the world either, the important point is that critical theorists acknowledge this. They argue that this recognition of the *practical* element of all theoretical ideas allows them to use theory to help bring about progressive social change.

DISCOURSE ANALYSIS AND CRITICAL GEOPOLITICS

Recently, a number of writers on international relations and political geography have turned to a particular set of approaches which focus on the *dis-*

courses of geopolitics and international relations. These discourses include those produced by politicians, diplomats, newspapers, firms: organizations and individuals involved in the practice of international relations. However, discourse analysts share the viewpoint of critical theorists on the practical status of all knowledge and discourses. Therefore their ideas apply also to the writings of academics and those claiming to be engaged in an 'objective and detached' interpretation of the activity of international relations. Writers who use the techniques of discourse analysis in the international relations field have classified their studies as 'critical geopolitics' to distinguish their approach from traditional realism.[16]

In this perspective, ideas expressed by practising politicians are not seen as false accounts of a true reality. Rather, they are seen as ideas which make certain things *become real*. To take a simple example, during what I have called the second cold war, Ronald Reagan referred to the Soviet Union as the 'evil empire'. From the viewpoint of discourse analysis, what is interesting is not whether this description of the USSR is true or false, but what the political *effects* are of using that kind of language in the conduct of foreign affairs. Some might say that this labelling of political opponents is mere rhetoric designed to make a political argument, and in one sense it is. But the point made by discourse analysts is that *all* writing and description involves rhetoric. If ideas are presented in a clear, rational and unemotional way, that does not somehow remove rhetoric from an account – clarity, rationality and lack of emotion are themselves rhetorical devices, chosen to achieve certain effects, such as promoting an aura of scientific authority.

During the cold war, the Soviet Union was regarded by Western governments as expansionist, militaristic and threatening to free-market capitalism and liberal democracy. The purpose of a discourse analytic approach to the cold war is not to try to decide whether that view was true, or even rationally held. Many discourse analysts would argue that such questions are unanswerable. What discourse analysis does is to examine the linguistic and symbolic means by which the Soviet Union was constructed and represented as expansionist, militaristic and so on. This means that critical geopolitics is concerned particularly with the 'texts' of international politics, rather than, for example, historical events, such as wars, in themselves. Indeed the idea that one could understand an event in itself without reference to the way it is presented rhetorically and discursively would not be accepted by those using the critical geopolitics approach.

A wide range of texts and other symbolic products and practices can provide the objects of investigation for critical geopolitics. I have already mentioned politicians' speeches in passing. The list could be extended to include policy documents, government reports, international treaties, journalism, novels, popular culture, military and 'patriotic' rituals, flags and (importantly, given the previous arguments about critical theory) academic writing and research. The process of analysis itself considers a number of aspects of texts, including their rhetoric, figures of speech, metaphors, forms of reasoning and symbolization and so on. What is stressed is the relationship between these linguistic or symbolic devices and political practices. How does a government justify a particular intervention? What kinds of language are used to represent come countries as friends and others as enemies? What sort of symbolic manoeuvre is happening when a government presents its interests as the 'national' interest?

According to Simon Dalby, one of the pioneers of critical geopolitics,

> doing these kinds of analyses subverts the discursive practices of conventional politics, calling into question all the silences and taken-for-granted constructions on which they are based. By refusing to accept reality as presented by the dominant discourses, numerous new ways of looking at politics are opened up. These challenge the conventional notions of both scholarship and political practice. Theory is not just a tool of analysis here, rather it too is the object of analysis, following the Foucaultian theme of asking questions about the production of questions. By refusing the . . . assumptions of clear lines of demarcation between theory and practice, theoreticians and practitioners, the social construction of practice in terms of theory is revealed and some possibilities for 'thinking otherwise' emerge; a process that conventional theorists find very unsettling indeed.[17]

In the Chapter 3 I showed how feminist challenges to the state were based in part on a critical understanding of the role of war and militarism in the process of state formation. This perspective is also a key emerging area of interest in critical geopolitics. Simon Dalby argues that the discourses which critical geopolitics analyses often depend heavily on gendered language and symbolism.[18] Feminist studies of geopolitics challenge many of the assumptions on which the practices of international politics are based. The discourse of 'national security', for example, is related to the existence of a feminine domestic sphere of hearth and home, which must be protected from external 'threats'. In the Gulf War of 1991, official coverage tended to ignore gender. The 'silences' in discourses are often as revealing as the statements themselves, and the Gulf War was in fact highly gendered[19]. Women suffered as victims, while the 'heroic aviators' of the United States airforce were represented as the epitome of masculinity. This discourse of masculinity was itself contradictory, involving simultaneously both extraordinary acts of violence such as the bombing of Baghdad, and cool 'rationality', with continual references in the media and official pronouncements to the 'clinical' and 'precise' quality of the attacks.

To show how this approach works in more detail, it will be helpful to consider an example. The example concerns the principal topic I have considered in this chapter – the relationship between the United States and the USSR. It comes from the work of Gearóid Ó Tuathail and John Agnew.[20] They argue that geopolitics should be reconceptualized in terms of discourse. They note the same significant link between material and discursive activities that I highlighted in Chapter 1:

> Geopolitics, some will argue, is, first and foremost, about *practice* and not discourse; it is about actions taken against other powers, about invasions, battles and the deployment of military force. Such practice is certainly geopolitical but it is only through discourse that the building up of a navy or the decision to invade a foreign country is made meaningful and justified. It is through discourse that leaders act, through the mobilization of certain simple geographical understandings that foreign-policy actions are explained and through ready-made geographically-infused reasoning that wars are rendered meaningful. How we understand and constitute our social world is through the socially structured use of language. Political speeches and the like afford us a means of recovering the self-understandings of influential actors in world politics. They help us understand the social construction of worlds and the role of geographical knowledge in that social construction.[21]

With this as their starting point, the authors go on to argue that there are four elements to their understanding of the relationship between geopolitics and discourse. First, they suggest, geopolitics as a political activity is not limited to a small group of specialized geopoliticians. Rather it is evident in all the ordinary debates about foreign policy conducted throughout the political system. Second, they argue that geopolitical reasoning is a practical process, which relies on 'unremarkable assumptions about places and their particular identities',[22] rather than an elaborate and formal process using specialist concepts. Third, they argue that geopolitical reasoning is related to the production of geographical knowledge in all its forms and throughout the social world.[23] Finally, they argue that the distribution of power between states means that certain, more powerful, states (such as the US) are able to influence disproportionately the production and character of discourses, and hence the wider understanding of geopolitical circumstances.

Having established the principles for their study, Ó Tuathail and Agnew proceed to show how the USSR was represented in discourse and 'scripted' by the United States. In doing so they focus on two texts produced at the start of the cold war by George Kennan, who worked in the US Embassy in Moscow and in the State Department in Washington. These were his 'Long Telegram' and the 'Mr X' article in *Foreign Affairs*. According to Ó Tuathail and Agnew, there are at least three significant discursive strategies pursued by Kennan in writing about the USSR. First, they suggest that Kennan represented the USSR as 'oriental'. This locates the USSR geographically as part of the Other (non-Western) world. Thus in the Long Telegram, Kennan writes that the Soviet government is pervaded by an 'atmosphere of oriental secretiveness and conspiracy'.[24] Second, they argue that Kennan constructed the USSR as sexually predatory, through his use of metaphors. For example, he referred to the Soviet leaders as 'frustrated' men who were subject to 'instinctive urges' and determined to 'charm' the West through Russia's 'primitive political vitality'. Third, Kennan's texts use the metaphor of the flood to refer to the communist threat to the West. The Soviet Union's 'political action is a fluid stream which moves constantly, wherever it is permitted to move, towards a given goal. Its main concern is to make sure that it has filled every nook and cranny available to it in the basin of world power.'

What the discourse in which Kennan's texts are located does is represent a place (the Soviet Union) and its relationships to other places (especially the 'Orient', Europe and the United States) in particular ways. It is through the language, symbolism and metaphors of such discourses that politicians (and the public) come to understand the place of 'their' country in the world and its relations to others. Because those understandings go on to shape practical policy activities, discourse helps to constitute political relations as well as describe them in specific ways.

High and low politics again

Traditional geopolitics has been concerned exclusively with high politics: wars, grand strategy and diplomacy. By focusing in part on the links between 'geopolitical reasoning' and popular culture and everyday understandings, critical geopolitics begins to break down the division between high and low politics. In the final chapter of the book, I want to develop this theme in more detail by turning to the political geographies of social movements.

Notes on Chapter 5

1 George F. Kennan, ('Mr X'), 'The Sources of Soviet Conduct', *Foreign Affairs* 25 (1947), pp. 566–82.

2 State of the Union address, 25 January 1980.

3 For a summary of the events involved, see Ronald Suny, 'Incomplete Revolution: National Movements and the Collapse of the Soviet Empire', *New Left Review* 189 (1991), pp. 111–25.

4 Francis Fukuyama, 'The End of History?' *The National Interest* (Summer, 1989). Fukuyama's arguments have been developed in his book: Francis Fukuyama, *The End of History and the Last Man* (New York, Free Press, 1992). See also the debate in *Political Geography* 12, 1 (1993).

5 Charles W. Kegley, Jr, and Gregory Raymond, *A Multipolar Peace? Great Power Politics in the Twenty-first Century* (New York, St Martin's Press, 1994), p. 5.

6 Peter Taylor, *Political Geography: World-economy, Nation-state and Locality* (London, Longman, 1989), p. 45.

7 *Op. cit.*, p. 46.

8 Halford Mackinder, 'The Geographical Pivot of History', *The Geographical Journal* 23 (1904), pp. 421–42; *Democratic Ideals and Reality* (London, Constable, 1919). Gearóid Ó Tuathail has pointed out that there was much more to Mackinder's life and work than the heartland theory for which he is best known. See Gearóid Ó Tuathail, 'Putting Mackinder in his Place', *Political Geography* 11 (1992), pp. 100–18.

9 Daniel S. Papp, *Contemporary International Relations: Frameworks for Understanding* (New York, Macmillan, 1991).

10 John Agnew and Stuart Corbridge, 'The New Geopolitics: The Dynamics of Geopolitical Disorder', in Ron Johnston and Peter Taylor, eds, *A World in Crisis: Geographical Perspectives* (Oxford, Blackwell, 1989), pp. 266–88.

11 *Op cit.*, pp. 268–9.

12 *Op cit.*, p. 269.

13 *Op. cit.*, pp. 279–80.

14 Simon Dalby, 'Critical Geopolitics: Discourse, Difference and Dissent', *Environment and Planning D: Society and Space* 9 (1991) pp. 261–83.

15 For a detailed critique of realism from the critical theory perspective, see Jim George, *Discourses of Global Politics: A Critical (Re)Introduction to International Relations* (Boulder, CO, Rienner, 1994).

16 *Op. cit.* See also: Dalby, 'Critical Geopolitics'; Simon Dalby, *Creating the Second Cold War: The Discourse of Politics* (London, Frances Pinter, 1990); Klaus-John Dodds, 'Geopolitics and Foreign Policy: Recent Developments in Anglo-American Political Geography and International Relations', *Progress in Human Geography* 18 (1994), pp. 186–208; Klaus-John Dodds and James Sidaway, 'Locating Critical Geopolitics', *Environment and Planning D: Society and Space* 12 (1994), pp. 515–24; Gearóid Ó Tuathail, 'Problematizing Geopolitics: Survey, Statesmanship and Strategy', *Transactions of the Institute of British Geographers* 19 (1994), pp. 259–72; Gearóid Ó Tuathail, '(Dis)placing Geopolitics: Writing on the Maps of Global Politics', *Environment and Planning D: Society and Space* 12 (1994) pp. 525–46; Gearóid Ó Tuathail and T. Luke, 'Present at (Dis)integration: Deterritorialization and Reterritorialization in the New Wor(l)d Order', *Annals of the Association of American Geographers* 84 (1994), pp. 381–98.

17 Dalby, 'Critical Geopolitics', p. 269.

18 Simon Dalby, 'Gender and Critical Geopolitics: Reading Security Discourse in the New World Disorder', *Environment and Planning D: Society and Space* 12 (1994), pp. 595–612.

19 Cynthia Enloe, *The Morning After: Sexual Politics at the End of the Cold War* (Berkeley, CA, University of California Press, 1993).

20 Gearóid Ó Tuathail and John Agnew, 'Geopolitics and Discourse: Practical Geopolitical Reasoning in American Foreign Policy', *Political Geography* 11 (1992), pp. 190–204.

21 *Op. cit.*, p. 191.

22 *Op. cit.*, p. 194.

23 In a fascinating study of the role of the media and popular culture in the cold war, Stephen J. Whitfield shows precisely how, in the United States, the fear of communism and of the Soviet Union was reproduced through film, television, literature and the press. See Stephen J. Whitfield, *The Culture of the Cold War* (Baltimore, MD, Johns Hopkins University Press, 1991).

24 All quotations from Kennan are cited in Ó Tuathail and Agnew, 'Geopolitics', pp. 199–201.

|6|

The geographies of social movements

Overview

This chapter looks at the nexus between formal and informal politics. Social movements are collective endeavours by groups of people in pursuit of political goals. While they often seek political change at least partly through the formal institutions of government and the state, they also operate outside of the arenas of formal politics. This chapter takes a broad definition of social movements which includes nationalist movements and labour movements as well as the so-called 'new social movements' of environmentalism, peace, feminism and anti-racism.

The chapter is organized into four sections. In the first, I introduce the concept of social movements and locate the study of them within the social sciences and human geography. The second section discusses the nature and characteristics of social movements. In the third part I turn to the issues surrounding the interpretation of social movements and consider how the study of social movements might be undertaken within the interpretative framework I have adopted in this book. The final, and longest, section looks at the relations between geography and social movements and includes three case studies: of the geographies of ethnic nationalist movements, labour movements and feminist movements respectively.

Introduction

Politics, I declared in Chapter 1, is first and foremost about people. In the following chapters people appeared in a variety of ways, as citizens, voters, colonial subjects and so on. On the whole, though, the perspective has been somewhat 'top-down' – seeing people in their relations to political institutions rather from the starting point of the institutions (such as the state). In this chapter I want to reverse that perspective and look at politics 'from the bottom up', by considering the geographies of social movements.

Unlike some of the other topics in the book, such as the state, there is no well-developed tradition within geography of studying social movements as a whole. Particular types of social movements have been the focus of study. Nationalist movements are the subject of a large literature within political geography, while Manuel Castells's concept of urban social movements provoked considerable debate and research among social and urban geographers during the 1980s. However, there have been few attempts to consider what geographers might have to offer in thinking about social movements in general. A short chapter like this cannot provide a full account of what such a geographical contribution to the topic might be, but I hope that it will provide an outline of some of the issues at stake.

The study of social movements straddles political geography on the one hand and social and cultural geography on the other. As their name suggests, their roots lie in social relations and interactions, but their objectives and modes of action are political. Within political geography, consideration of the relationships between individuals and the political sphere has often focused on elections (see Chapter 3) and, more recently, on citizenship. This emphasis to some extent reinforces the liberal democratic view of legitimate politics as a formalized and limited relationship between individual and state. A focus on social movements, by contrast, challenges these limits to what is considered legitimate. Most social movements question many of the taken-for-granted assumptions about the present distribution of social and political power and the presumed legitimacy of existing political institutions and practices. In some cases, social movements are concerned with trying to establish liberal political and civil rights (for example, those which led to the downfall of the state socialist governments of the former Soviet Union and Eastern Europe). In others, they may be questioning the gap between the freedoms and entitlements formally offered by the existing system and the perceived failure of all groups to benefit from them equally in practice. More radically still, some social movements seek to challenge the whole basis on which conventional politics is practised.

According to the sociologist Anthony Giddens, the study of social movements has been neglected in the social sciences as a whole. As I have suggested, this neglect has certainly not been total within geography, but there have been few systematic accounts of the political geography of social movements. This is something of a paradox, since many geographers have in recent years espoused the notion, originally put forward by Karl Marx, that people 'make history, but not under conditions of their own choosing'. While much work has been done on the 'conditions not of their own choosing', relatively little has been done on the making of history by people. While by no means representing the whole of that story, social movements are clearly one of the most important mechanisms which people have to enable them to make history. Before we consider how that making of history both depends on and creates geographies, we need to look in a little bit more detail at the character of the social movements involved.

What are social movements?

Collective action and political strategies

Social movements may be defined as groups of people acting collectively in pursuit of shared goals which include, or require, social and/or political change. They are thus different both from social clubs or voluntary organizations which do not have political goals, and from religious movements except in so far as these may have earthly political objectives. Giddens cites Herbert Blumer's definition of social movements as 'collective enterprises to establish a new order of life'.[1] It is important to recognize both elements. The enterprise is collective because it involves people acting together, not just as individuals, and because it is aimed at producing social and political change (a new order of life), rather than, say, social companionship, sporting prowess or prize vegetables. Having said that, social movements vary in the extent of change which is sought and in the elements of society in which change is thought necessary. A revolutionary movement may seek the wholesale overthrow of the existing social order, while a movement for voting reform may be concerned more narrowly with the extension of political rights within the existing political framework.

The desire to achieve change reveals another characteristic of social movements: they are oppositional. That is, they are opposed to one or more elements of the existing social and political order. This further implies that they are in conflict with other groups or institutions in society who wish to preserve those elements. It is also usually the case that this process of conflict reflects and draws on a more general conflict of interest in the wider society. Some social movements are thus regarded as 'single issue' because they focus on and organize around one axis of division within society. By contrast, established political parties are sometimes seen as different from social movements in making a broad appeal across all issues (not least because a government must attempt to deal with the whole variety of policy questions in society, rather than just those which touch on one issue). However, it would be a mistake to draw too rigid a distinction here. Social movements can become political parties, and some political parties do appeal to very particular interest groups. Moreover, the beliefs of social movements which seem to be 'single issue' can in fact involve a broad view of society as a whole and form the basis for policy proposals in a wide range of fields. A good example is the women's movement, which may appear at first sight to be a single issue ('rights for women') movement, but which contains within it a variety of traditions of feminist thought which involve a critical perspective on many aspects of social and political life: such as the provision of public services, military activity and the organization of the state itself, social welfare and economic and industrial policies.

I have been arguing throughout the book that the practice of politics involves strategies. Often the political implications, and the political changes which arise as a result of strategic activity are different from those which were intended during its development. This is clearly true to some extent of social movements – their impact is never wholly predictable or controllable in advance. However, they do differ from a number of the other topics we have looked at inasmuch as they can develop quite deliberate and purposeful strategies from an early stage, which are intended to produce political effects. Social movements are thus strategic in an unusually strong sense. This results in another key feature: one which Giddens refers to as 'the reflexive monitoring of action'. According to Giddens, 'reflexive monitoring' is a permanent feature of social life in general. What he means is that we all, in our daily lives, observe ourselves and our actions, and adjust our future actions in the light of the knowledge we gain of ourselves, others and our surroundings. For the most part, he argues, we do not think about this very much, we just do it. One of the distinguishing features of social movements is that the process of reflexive self-regulation is more explicit and developed than in daily life. The participants in social movements want to have a particular kind of effect on the wider society, and this implies a deliberate attempt to steer the activities of the movement in the light of evaluations of its past successes and failures.

Social movements and organizations

According to Giddens, social movements are to be distinguished from organizations, partly on grounds of their relationships to space. Thus organizations, he suggests, commonly operate within fixed 'locales'. What he means by this is that specific types of organizations such as the army, businesses, hospitals and prisons have specific settings which to some extent structure their activities: the barracks in the case of the army, factories and offices for businesses, wards, clinics and operating theatres for hospitals, and cells and guardhouses in prisons. It is perhaps not possible to draw the line precisely, and some modern organizations are beginning to find ways of escaping the geographical fixity implied by 'locales'. On the other hand it does seem that social movements are less institutionalized than many organizations: they seem to be less organized! However, social movements can develop organizational attributes and can become institutionalized. Thus the labour movement, which Giddens sees as the archetypal social movement, is, in many countries, made up to a large extent of formal organizations: the trade unions. Similarly, although it is more diffuse, the women's movement certainly has its organizations, whether they are fairly informal local protest groups, or major institutions like the National Organization of Women in the United States.

In Chapter 1 I outlined the distinction between formal politics, involving institutionalized political organizations, governments and so on, and informal politics, which is much more concerned with the power relations of everyday life. Social movements blur these distinctions. Often they begin through an awareness based in problems of everyday life, but they can become major political institutions (such as political parties) in their own right. Moreover, they also provide a key link between formal and informal politics by enabling issues from the informal arena to be placed on the formal political agenda. At the same time, they involve direct participation of ordinary people in the political process.

Modern political parties certainly share some of these characteristics. They usually are, or aspire to be, mass organizations, and many of them developed from social movements in the first place. Thus the British Labour Party was established originally as the parliamentary branch of the labour movement. On the other hand, once political parties become serious candidates for elected office they tend to lose some of the clear characteristics of social movements, and become, in some ways, a part of the state. As part of the existing 'system' they are likely then to be a focus of opposition for social movements operating outside the parliamentary arena, who are working to try to change 'the system' rather than to join it.

Human geographers, and other social scientists, often point out that social change is the product of social struggle and that, historically, political measures which are seen as 'progressive' are rarely handed down 'from the top' but have to be struggled for from below. What the study of social movements provides is an understanding of how such struggles actually take place in more concrete terms. Struggle and resistance can take many forms, from minor graffiti to revolution. Struggles usually result in major social change only when they are channelled through social movements. It is possible to make a difference to the way society works, but it is very difficult to do it on your own!

The varieties of social movements

Social movements are of many kinds, and have many different foci. One of the first and historically most influential social movements was the labour movement, based on opposition among workers to the conditions of work which developed during the nineteenth century with the rise of industrial production and the capitalist economic system. Also of great importance during the twentieth century have been nationalist movements of various sorts, including the kinds of anti-colonial nationalism I discussed in Chapter 4. One group, the so-called 'new social movements', are commonly counterposed to 'traditional' labour and nationalist movements. The new social movements include the women's movement, the environmental movement, the anti-racist movement and the peace movement.

Since the late 1970s an important focus for geographers studying social movements has been the concept of 'urban social movements' developed by Manuel Castells.[2] Castells based his ideas initially in the Marxist concept of the capitalist mode of production. He argued that the city was defined as the arena of social reproduction of the labour force. With the development of capitalism more and more of the means of social reproduction, such as housing, health care and education had come to be provided by the state. The city was thus also the site of struggle and conflict over the state provision of these services, and urban governments were frequently the focus of campaigns by urban social movements to improve the provision of public services. In his later work, Castells broadened the focus away from the narrow (though important) issue of the reproduction of the labour force to include the new social movements mentioned above.

The term 'new social movements' is commonly used to identify movements which rose to particular prominence in the 1960s and 1970s. In many ways, however, it is something of a misnomer. Many participants in the 'new social movements' have pointed out that the conflicts and struggles on which they are based have a long history, stretching back to the nineteenth century and beyond. In addition, there are major differences between the new social movements, although according to Camilleri and Falk they are rooted in three aspects of modernity. These are: the breakdown of traditional communities, especially following the development of large cities; the rapid growth of technological development and the consequent threats to environmental and military stability; and the failure of the state to resolve the contradictions between economic growth and development and their social, cultural and environmental effects.[3] These broad social changes certainly provide important contexts for the development of social movements in the late twentieth century. However, specific movements involve responses to quite particular elements of modern life, and it is to these I now wish to turn.

Understanding social movements

Existing approaches

Approaches to understanding social movements commonly fall into two large groups: those which stress the objective conditions which give rise to social movements, and those which concentrate on the subjective experiences which prompt people to join and participate in them. There are problems with this dichotomy, but it is a useful starting point for the arguments which follow. First, some approaches focus on the objective conditions which give rise to social movements. For example, if one ethnic group is systematically disadvantaged economically in comparison to another, an ethnic nationalist movement may arise. Systematic inequalities which promote the formation of social movements are sometimes analysed as the product of deliberate discrimination and sometimes as the product of the structural properties of the social system. An example of the former might be the anti-apartheid movements in South Africa, with apartheid being viewed as a deliberate and conscious strategy on the part of the white minority to control society in their interests and against the interests of the black majority. An example of the latter is a version of the classical Marxist approach to the analysis of classes in capitalism, which sees the disadvantages suffered by the working class relative to the bourgeoisie as the product of the 'logic' of the process of capital accumulation. In this view, the rise of the labour movement stems from the objectively opposed interests of the working class and the bourgeoisie.

Second, there are those approaches which focus on the development of a subjective sense of group belonging or group disadvantage. In the analysis of nationalist movements, for example, this perspective would stress the importance of a sense of belonging to a national community. In the labour movement example, the emphasis would be on the emergence of class consciousness.

Both of these broad perspectives have their merits. Clearly the social and economic conditions in which social movements develop and operate are likely to have important influences on their strategies and success. Equally, as political movements comprised of committed individuals, the politics of social movements must be seen in part as the consequence of the views, emotions and perceptions of the people who make them. However, both perspectives have their limits. If we place particular stress on the objective conditions, it is difficult to explain why social movements arise in some situations, but not in others which appear, objectively, to be as appropriate. Similarly, it is difficult to account for the development of similar kinds of movements in very different circumstances. If nationalism is produced by objective economic disadvantage, how can we explain the greater strength of ethnic nationalism in Scotland than in Wales when Scotland has the stronger economy? Indeed there seems to be no consistent pattern between socio-economic conditions and the rise of nationalist movements. Economic problems seem to translate into socialist revolutions in some contexts and nationalist uprisings in others.

The second, subjective perspective seems at first glance to offer the solution to this conundrum. Perhaps different responses to similar circumstances are the result of different people being involved with different ideas and perceptions, interpreting their situations in different ways. The problem with this is that it begs the question of how people's ideas and perceptions are formed in the first place. Since it seems likely that they will be heavily influenced by the circumstances in which they develop, we are back where we started.

To some extent the question may be resolved by combining the insights of both points of view. Yes, social and economic circumstances are important, but while they influence the development of consciousness on the one hand, they are also interpreted by consciousness on the other. This interaction between social conditions and human perceptions is important. However, I want to develop the argument in a slightly different direction, by focusing on what seems to be 'political' about social movements, in the sense which I discussed politics in Chapter 1. Stressing socio-economic conditions on the one hand or individual consciousness on the other to some extent neglects the political nature of social movements. By this I mean that we need to focus on how and why particular human feelings or group attachments become mobilized in a political movement and how the contexts in which social movements are used by them in developing political strategies. This approach is the one which fits in with the interpretative framework I have been using.

Social difference and political identities

Many social movements, though not all, are associated directly with the personal identities of those involved and the politicization of those identities. For example, feminism involves a politicization of women's identities as women, social movements of black people work because their participants are conscious of a particular ethnic identity and so on. With some movements, the link with personal identity is more diffuse. For example, the environmental movement is not obviously organized around an element of personal identity. On the contrary, in many ways it tries to appeal to a shared sense of humanity, and seeks to be universal. However, although it is less stark, there are notions of identity here too, in so far as environmentalists challenge the basis of industrial society and thus the sources of people's identities as producers and, especially, consumers.

The tension between human universalism on the one hand and a stress on different identities on the other is discussed in detail in the work of Iris Marion Young. She outlines two contrasting approaches to dealing with the problems of inequality and social oppression. These 'competing paradigms of liberation'[4] are an 'ideal of assimilation' and an 'ideal of diversity'. The ideal of assimilation argues that liberation from oppression will be achieved when social differences cease to have political significance. Thus, human beings could be divided into groups according to the colours of their eyes, but no society or political system in the world makes any distinction at all (formally or informally) between people on the basis of eye colour. Drawing on the writing of Richard Wassertrom, Young outlines the assimilationist perspective as follows:

> A truly nonracist, nonsexist society . . . would be one in which the race or sex of an individual would be the functional equivalent of eye color in our society today. While physiological differences in skin colour or genitals would remain, they would have no significance for a person's sense of identity or how others regard him or her. No political rights or obligations would be connected to race or sex, and no important institutional benefits would be associated with either. People would see no reason to consider race or gender in policy or everyday inter-actions. In such a society, social group differences would have ceased to exist.[5]

Young accepts that the ideal of assimilation has been very important in politics by stressing, among other things, 'the equal moral worth of all persons, and thus the right of all to participate and be included in all institutions and positions of power'.[6] It also challenges the still widespread popular assumptions that certain groups in society are inherently inferior. However, Young herself prefers the alternative 'ideal of diversity'. She notes that although the assimilationist position has its attractions, it remains something of a distant utopia, and that in the present climate social groups have turned to the distinctive aspects of their identities as sources of strength. The ideal of diversity stresses respect for difference rather than its erasure, and also insists that certain social differences require different treatment (for example, treating men and women exactly the same might result in neglecting women's particular needs during pregnancy and childbirth).

Writing of the situation in the United States, Young outlines the move towards political movements based on social difference. This has involved a variety of social movements:

- the 'Black Power' movement, which argued, for example, that 'Black English is English differently constructed, not bad English' while 'Afro-American hairstyles pronounced themselves differently stylish, not less stylish';[7]

- the 'Red Power' movement asserting the right of native Americans to self-determination and distinctive cultural practices;
- the gay and lesbian liberation movements, which reject the dominant assumption that a heterosexual lifestyle is 'normal' or 'more healthy', and stress pride in gay and lesbian sexual identities and sexualities;
- the feminist movements, many of which have turned away from the stress on improving the position of women in existing institutions and seek to challenge the patriarchal assumptions around which the institutions are built. They also stress the positive aspects of qualities understood as feminine, such as caring, nurturing and co-operation.

For Young, the importance of a continued emphasis on social difference arises because of both continuing oppression of some groups by others, and the political and cultural strength which comes from group identity. Difference is thus a positive aspect of society and need not be the basis of systematic discrimination. There are, however, dangers in this approach. One key objection to stressing social difference is that it is essentialist. This means that identity differences are seen as inherent and enduring features of society. Stressing the differences between men and women and between masculinity and femininity, it is suggested, means arguing that there is some essence or ultimate core of femininity and masculinity, which makes men and women inescapably different from each other. For feminists, who regard masculinity as oppressive, therefore, essentialism might imply an uncomfortable choice between permanent oppression or separatism. Gillian Rose has identified this as part of an ambivalence in much feminist thinking.[8] On the one hand feminists want to assert that women and men are alike insofar as feminists wish women to gain access to rights and privileges currently accorded to men. On the other they want to assert that women are different from men in so far as masculine traits are seen as socially oppressive. A similar ambivalence between universalism and essentialism is probably characteristic of most politics of identity.

Iris Marion Young insists that emphasizing difference does not mean adopting an essentialist notion of identity. She proposes a notion of difference which defines it in terms of the relations between groups, rather than as essential characteristics of groups. This means that the process of group formation is not a rigid and objective one, in which individuals can be assigned to groups on the basis of established and enduring identities. Social movements based on identity groupings are thus porous, not exclusive:

> Membership in a social group is a function not of satisfying some objective criteria, but of a subjective affirmation of affinity with that group, the affirmation of that affinity by other members of the group, and the attribution of membership in that group by persons identifying with other groups. Group identity is constructed from a flowing process in which individuals identify themselves and others in terms of groups and thus group identity itself flows and shifts with changes in social process.[9]

In addition, it follows from Young's conception of difference and identity, that all individuals have multiple sources of identity. We are not just men and women but also working class and middle class, able bodied and disabled, healthy and sick, rich and poor and members of different ethnic and national groups. Each individual is at the centre of a web or network of potential multiple identities. I say potential, because not all sources of social difference form equal elements in the personal identities of individuals. One person may be particularly conscious of their gender, while another stresses their ethnicity. In principle, however, identities are multiple and even fragmented, which further undermines the possibility of seeing certain characteristics as essential.

This raises a further question, however. If identities are multiple, then why do particular identities form the basis of political movements and not others? This happens because different identities are politicized in different ways and at different times and places. To consider this issue we need to move from the social bases of social movements to their more explicitly political characteristics.

Discourses and resources

As I have suggested, one of the difficulties facing the analysis of social movements is accounting for the rise of specific social movements at particular times and places. I noted that however significant the wider conditions in which social movements develop may be, they are not sufficient to explain the temporal and spatial variation in the development of social movements of different types. One element in this is provided by Young's account of social differentiation. However, it is necessary to go further and to ask how and why particular social differences form into social movements while others do not.

In line with the interpretative framework of the book, I want to focus on two aspects of the politicization of social identities: discourses and resources. First, the development of a social difference into a social movement involves the discursive construction of that difference as being of political importance. Second, the ability of a social movement to capitalize on that politicization will depend in part on the mixture of resources which it is able to mobilize.

The emergence of social movements is never a smooth process. However coherent a vision the movement may eventually develop, it is rarely the case that it is thought out and expressed coherently and in full at the beginning of the movement's existence. When Giddens refers to the reflexive self-regulation of social movements, he means, among other things, their capacity to adapt and refine their strategies, objectives and methods over time.

Nevertheless, for a movement to begin this process of development it must be identified, discursively, as a political phenomenon. The origins of this discursive construction vary from movement to movement. For example, many writers have stressed the role of ethnic élites, or the ethnic intelligentsia in the formation of nationalist movements. Such élites, it is argued, have access to, and responsibility for, key symbols of the national culture with which to generate support for the movement as a political cause. In many cases, organizing to protest at a very specific and immediate political issue can provide the spark which identifies a certain set of social relations as politically charged. Arguably the campaigns for women's suffrage in the early years of the twentieth century fulfilled this function in the women's movement. Once the initial limited objectives have been achieved, there is both momentum and organization through which the movement may be widened and deepened.

Whatever the trigger which gets things moving, discourses of and about the movement are central to keeping it going. These can take many forms. They include, for example, narratives about the history of the movement itself, with reference to its 'great thinkers', important activists, tragic defeats and glorious victories. In addition, there are political discourses which represent the movement as a struggle against (usually) oppression or discrimination. Thus the social relations around which the movement is organized are themselves in part constructed through discourses. Labour movements generate particular discourses about labour relations, around rhetorical figures such as 'the bosses' and 'the workers', the environmental movement uses metaphors such as 'spaceship earth' and describes the biosphere as 'fragile' or 'vulnerable'. Nationalist movements often rely heavily on 'patriotic' symbols of the national culture.

It is important to stress that these discourses and rhetorical strategies are not simple propaganda produced by a political movement to try to win over support from without. They are also centrally concerned with developing and retaining the always incomplete and unstable coherence of the movement itself. Because political identities are themselves unstable and prone to fluctuate, at times social movements have to work hard to keep a unified sense of purpose and direction. Moreover, just as Young's concept of social difference works through relations, rather than essentialized characteristics, so the discursive representation of difference as political is also a relational process. What this means is that social movements do not produce their own discourses independently of other groups and social processes surrounding them. A group or movement can be marked as distinct or political or both in the discourses of others (including its political opponents) as well as in the discourses it produces itself. Indeed, it is probably a mistake to think of movements having their own discourses. They will certainly engage in discursive practices, but these will almost certainly draw on, engage with or contest wider discourses which are produced through the interaction of a variety of individuals, groups and agencies. Discourse is rarely the property of a single agent. Since discourse is about communication, it is likely that any one group must use and engage with existing discursive formations in order to communicate. In doing so it will change their meaning, whether consciously or otherwise; but it is only by accepting at least a minimum shared agenda that others will be able to understand it. Nationalism is a good example. However particular the details of a given national consciousness, all nationalisms appeal to the wider (and widely shared) belief in the idea of nations and in the benefits of national self-determination. Specific nationalisms and their discourses thus 'fit in' with a general discourse of nationalism. One of the criteria for success for social movements is their ability to generalize their discursive understandings so that they begin to seem like 'common sense'.

Such success varies markedly between social movements. The reasons for success and failure are complex and varied. One way to begin understanding the problem is to think about the resources which movements are able to mobilize in support of their cause. 'Resource mobilization theory' is a particular theoretical perspective which seeks to account for the success of social movements in terms of their use of resources.[10] To summarize, resource mobilization theory argues that social movements arise from deliberate organizational strategies. These strategies depend on the availability of resources to the movement. Resources may be of all kinds, including money and material resources, symbolic resources, organization capacity and people's time and commitment. In this, the discursive formations I discussed above may themselves be seen as forms of symbolic resources.

The original formulations of the resource mobilization approach were grounded in 'rational choice theory', which argues that people act and develop strategies on the basis of rational calculations about the costs, benefits and likely outcomes of different courses of action. On this reading, resource mobilization theory sees social movements as the products of rational action in which resources are used collectively and deliberately in an instrumental and utilitarian fashion to achieve specific goals. Many writers object to rational choice theory because it assumes that people do in fact act according to rational calculations of their self-interests, when in fact much human behaviour is impulsive and emotional, and people rarely make conscious and instrumental calculations about the relative costs and benefits of different courses of action.

However, some of the important insights of resource mobilization theory do not depend on the rational choice perspective. In line with the framework outlined in Chapter 1 we can also draw on the work of Giddens to provide further help. According to Giddens, human beings know quite a lot about the circumstances in which they act, but their knowledge is far from complete. This means that their actions have unintended and unpredictable consequences. However the success and impact of action (individual or collective) is still seen as dependent in part on the resources which people can mobilize in the pursuit of particular goals. In other words, access to and the use of key resources, such as people's time, symbols, discourses and money are crucial to explaining the success or failure of social movements, but that does not mean that their use can necessarily be planned rationally in advance.

A combined approach

It seems that in order to produce a full account of the development and effects of social movements we need to understand both the meanings they carry and the motives of their members *and* the means of operation, their resources and organizational capacities. It is important to consider the division of the social world into social groups which are both the target of social discrimination and the source of emotional, cultural and political strength. However, not all such groups form the basis of social movements and not all social movements are equally successful. To explain the rise of social movements and their pattern of success we need to pay attention to the process of their politicization through discourse and the resources they are able to mobilize. The political outcomes of these processes are intimately connected with geography, as we shall see in the next section.

The approach I have outlined can help to guide analysis and interpretation, but it should not be considered as a rigid theory of social movement development. Although they do share some common characteristics, modern social movements are highly diverse. I have concentrated on those which are based on social group identities, such as gender, class or ethnicity. However, there are important social movements which do not correspond quite so tightly to social identity, or social groups. The environmental and peace movements, for example, make their appeal on the basis of what are argued to be universal human values, such as the survival of the human species in the face of environmental degradation or the threat of nuclear war. Support for these movements does not come because of the oppression of a particular social group, but reflects what are felt by the participants to be general social concerns.

This does not mean, however, that these movements have support which cuts across all social groups. Membership of both the peace and environmental movements, for example, is strongest among middle-class social groups. Although these movements are not 'about' issues of group identity, they reflect social concerns which are more prominent and more politicized in some social groups than in others. With some modifications, therefore, the approach I have outlined, with its emphasis on the processes of politicization, discourse production and resource mobilization could reasonably be applied to these social movements too.

I noted above that social movements were not the same as organizations.

As we have seen, however, they are (to a degree) organized. But this should not lead to the assumption that social movements are tightly knit, homogeneous phenomena. On the contrary, in so far as they are organized, they often consist of many different organizations, each with a slightly different appeal, perspective or strategy. The environmental movement, for example, contains a wide variety of organizations, including campaigning groups, political parties, direct action groups, newspapers and magazines and a variety of informal groupings. Like the social groups on which many of them are based, social movements are much more porous than other forms of social collectivities such as firms.

The geographies of social movements

In this section I want to consider the relationships between geography and social movements. To begin I will briefly outline some of the connections I want to emphasize. These will then be illustrated in a selection of short examples: nationalism, trade unionism and feminism.

Space, place and social movements

Geography is important to the formation and activities of social movements for a number of reasons:

1 There are geographical differences and variations in the development of social groups and group identities. This may be partly as a result of the rise specifically of identities which are associated with particular places, sites, or use of space, and partly because the formation of identity is always conditioned by the geographical context in which it occurs.
2 There are geographical variations in the distribution of the resources which groups mobilize in trying to achieve their goals.
3 There are geographical variations in the distribution of other social movements, political institutions, economic circumstances and cultural understandings which influence the impact that particular social movements have in particular places.
4 Different social movements operate (or aspire to operate) at different geographical scales. Some may be highly localized, such as a regional separatist movement. Others may operate nationally, such as a civil rights campaign; at a continental scale (e.g. European Nuclear Disarmament (END)); or aspire to be global (some forms of feminism, environmental-

ism). Geographical scale has an impact both on the objectives of the social movement as a political entity and on its organizational requirements and capacities. Very few, if any, single political organizations are truly global in scope, but several broad social movements have had at least some impact in most parts of the world.

These general principles suggest that all social movements will have a geography. They will operate at a particular geographical scale or combination of scales, and are likely to be unevenly developed with stronger and more successful elements in some places than in others. Moreover, social movements are also constituted geographically; geographical variations have an impact on how they arise and develop. Here are three examples of this.

First there is the question of the geographical context in which social movements develop. Geographical contexts are important because they provide the resources through which social movements develop. 'Context' in this case should not be understood as necessarily implying a setting which is highly localized, although this may be the case. Many resources are available across a wide spatial field. However, none are available universally and entirely evenly. Access to resources depends in part on status (entitlement) but also on socio-spatial context. Giddens's concept of 'locale' may be helpful here. For Giddens, the notion of locale refers to the setting for social action which furnishes resources on which people draw in conducting their relations with others. He suggests that social movements do not characteristically operate in *fixed* locales, in the same way as many organizations, but this does not mean that locales are unimportant in the development of social movements. Moreover, 'locales do not have to be local'.[11] To take a simple example, the distribution of skills associated with publicity and fundraising is geographically uneven. Where a social movement can gain access to such skills it may be able to mobilize them in developing its activities. Where it cannot (and other things being equal) it may find it is relatively weaker. Similarly some discourses make sense to people in some places but not in others. An environmental discourse about the threats to ecological stability posed by the practice of Western science will make little sense in communities where Western science is unknown.

Following from this and secondly, there are marked variations in the characteristics of social movements between the so-called First, Second and Third Worlds. As Graham Smith points out, the 'new' social movements which developed in the West during the 1960s and 1970s do not have direct counterparts in the Second World of former state socialist countries in Eastern Europe and the former Soviet Union. In these countries the strong social movements of the late 1980s and early 1990s were developed specifically as challenges to the state socialist systems. Part of their objective was to create the kind of public arena and civil society missing in the context of state control of the economy and society. In the West, it was precisely the presence of a public arena which provided a space within which new social movements could develop. In the Third World social movements have been influenced by the context of colonization and de-colonization. Initially social movements developed around the national liberation struggles in colonized areas. Since formal independence many social movements have been concerned with the problems of poverty and the inability of the state to fulfil promises to produce rapid development. In addition, however, there are ethnic and other tensions which have produced social movements in the South which are not *directly* attributable to the impact of colonialism or continuing informal imperialism.

Finally, some social movements are organized around what are commonly regarded as geographical questions or issues. The environmental movement has often been of particular interest to geographers because it is concerned with something which geographers often study: environmental change and environmental problems. Nationalist and regionalist movements have also been the focus of much geographical research since they are explicitly concerned with the question of territory. More recently, geographers have considered the implicit geography in the concerns of a variety of other social movements, such as the feminist questioning of the gendered division of space into public and private spheres.

The geographies of ethnic nationalism[12]

In 1975 Tom Nairn identified nationalism as the 'modern Janus'.[13] This reference to the two-faced Roman god is common in writing on nationalism and attempts to capture something of the complexity and even contradictions of the idea. Nationalism is seen as facing two ways because it has both emancipatory and repressive elements. On one hand it has been the motive for struggles of liberation from colonial oppression, while on the other, it has motivated extreme hatreds culminating in genocide (most starkly during the Second World War).

In addition to this political variability, there are very marked geographical and historical variations in nationalist movements and conflicts. The Indian nationalism which led to independence was different from that which has generated conflicts between ethnic and religious groups in modern-day India. The German nationalism mobilized by the fascist Nazi party in the 1930s is different from the patriotism of the democratic government of 1990s Germany. The left-of-centre nationalism of the Scottish National Party or Plaid Cymru (Party of Wales) is different from the exclusionary, xenophobic nationalisms which are prising apart the communities of the former Yugoslavia. All of these phenomena are called nationalism. What is common to them, that they can share the same label?

Nationalism appeals to the ideas of the nation and of national identity (see below). Nationalists argue that there should be a congruence, in terms of territory and boundaries, between a nation and its political community. In the modern world, the political community means, effectively, the state. This desire for complete overlap between nation and state is widely accepted and taken for granted (although that does not mean that it is automatically a good thing) and it gives rise to the concept of the nation state in which nation and state are supposedly united. The aspiration is so widespread that the term 'nation state' is very commonly used to refer to states, even where these are not at all congruent with national groupings.

NATION AND NATIONAL IDENTITY

Nationalism is a political movement or set of political ideas based on national identity. Although it has fluctuated in intensity, it has been one of the strongest forms of political identity in the twentieth century. This strength has led some writers to attribute feelings of national identity to some kind of genetic, biological or instinctive process. One difficulty with this perspective is that it makes it very difficult to explain the dramatic variations in national feeling and nationalist activity. If 'human nature' is the cause, then how can we account for some humans who will happily kill or die for nationalist sentiment and others who express only weak national identity, or none at all? One alternative is that which I outlined above in discussing social difference and political identity. National identity is one of the many possible social identities which we have. It is based on the formation of a social group (the nation) which is differentiated from other groups (other nations) and from other sources of identity (gender, religion, sexuality and so on).

One important and influential form of nationalism is ethnic nationalism. In ethnic nationalism, the nation is defined around an ethnic group. The term 'ethnic' carries connotations of kinship and common genetic ancestry. These connotations are very important in the discourse of national identity, but it is important to emphasize that they are discursive constructions. Of course some members of a particular ethnic group are related to one another genetically, but the idea that all the members of the group are related is a myth which helps to define the group culturally, but which need not (and almost certainly does not) have any real biological basis. According to Anthony Smith there are six characteristics which define an ethnic community:

1 a collective proper name
2 a myth of common ancestry
3 shared historical memories

4 one or more differentiating elements of common culture
5 an association with a specific 'homeland'
6 a sense of solidarity for significant sectors of the population.[14]

To repeat, from the perspective I have adopted the important thing here is that these characteristics are all discursively constructed. This does not mean that they are 'mere' myths, because discourses can have extraordinary political power. Nationalists often argue that nations are defined around a central core identity, and in one sense they are. The focus of ethnic nationalism is the (discursively constructed) sense of ethnic identity. However, following Iris Marion Young's arguments about social group differences, the definition and construction of a sense of ethnic identity comes about through marking differences between different ethnic groups. Scottish culture is Scottish because it can be defined as different from Welsh or English culture. Scottish nationalists may well disagree strongly about what 'Scottishness' is, but all would define it as distinct from 'Englishness'.

Anthony Smith argues that national identities develop around ethnic identities, given particular social, economic and political changes. He suggests that for an ethnic group to become a nation there must be a stronger, more physical and immediate connection between the group and 'its' territory, and there must be not just common cultural 'markers' but also a common public culture. In short, nations involve shared practices as well as shared myths and memories. Nevertheless nations too remain in part symbolic and discursive constructions. Benedict Anderson argues that they are 'imagined communities', because while they provide a strong sense of community, that community feeling is secured in the imaginations of its members, since most members of a nation will never meet most of their co-nationals.

NATIONALISM AS A SOCIAL MOVEMENT

Ethnic nationalism as a social movement marks a transition from the existence of an ethnic and national group to a political process and political activity organized around it.[15] National*ism* involves the development and pursuit of political strategies through the use of resources. The usual goal of nationalism is the establishment of a political community (usually a state) with the same territory as the nation, so that the nation can achieve political sovereignty and control over its own affairs. This frequently means seeking political independence from an existing state which is seen as dominated by another nation. 'Irredentist' nationalism represents a special case in which a national minority on the border of one state seeks to join with co-nationals in a neighbouring state.

Many writers on nationalism have stressed the social and economic conditions which can promote the development of nationalism. Nationalist movements are subject to significant geographical variation in their development, intensity and success. Geographers and others have often sought to explain such variation in terms of the uneven spread of social and economic processes. Regions that remain peripheral to economic growth, or excluded and remote from the sources of, for example, state power, may develop nationalist movements. While such preconditions are very important, there is no universal rule about which kinds of economic, social or political problems will produce nationalist responses. In some cases, nationalism develops in relatively rich areas. In the former Soviet Union, for example, nationalism was a stronger force in the Baltic republics than in Soviet Central Asia, although the Asian republics were poorer economically. In other areas the pattern may be reversed. In other words, a whole variety of circumstances *can* provide a foundation for nationalism, but it is impossible to predict on the basis of such circumstances alone which *will*.

If we think of nationalism as political strategy, however, the picture becomes a little clearer. Once nationalism is understood as a political project which is pursued by certain individuals and social groups within the nation on the basis of the resources they are able to mobilize, we can begin to explain its emergence and geography.

One line of reasoning stresses the role of ethnic élites in this process. The 'cultural intelligentsia' within a nation can be important, because they often control, or at least have good access to, resources for communication and political mobilization. They are likely to be well educated, articulate, and familiar with the sources of cultural identity around which the ethnic group and the nation is discursively constructed. Furthermore, they may well have particular vested interests in achieving national autonomy, since it is the ethnic élite which would be likely to staff any new state apparatus, and benefit particularly from new sources of economic growth. For example, the greater development of ethnic élites in the Baltic republics of Latvia, Lithuania and Estonia helps to explain why nationalism developed there much more strongly than in Soviet Central Asia. The argument about resources is also important in this case. Because the Baltic republics had experienced a period of political independence between the two world wars, nationalists had particular resources on which to draw in making their case in the forms of historical memories, and a discourse of lost statehood. Such resources were not available in the same way in other parts of the former Soviet Union.

Nationalist strategies can also involve what Hobsbawm and Ranger call the 'invention of tradition'.[16] Nationalism involves a discourse of antiquity. It argues that nations have ancient networks of kinship and cultural belonging. In practice most nations are modern phenomena and in many cases what appear to be ancient cultural traditions are actually remarkably recent arrivals. As Hugh Trevor-Roper shows in his discussion of Scotland, many of the distinctive markers of 'Scottishness' (to outsiders as much as to Scots) are quite modern inventions, including kilts and bagpipes. Moreover they were often developed after the Act of Union which joined Scotland politically to England, and form in part a protest against it. While there is very much more to Scottish nationalism than this, the example serves to illustrate how significant elements of nationalist discourse can be constructed quite deliberately as part of particular political strategies.

The geographies of labour movements

CLASS, IDENTITY AND TRADE UNIONISM

My second example of social movements is the geographies of labour movements. Trade unions, which make up labour movements, are collective organizations of workers acting together to further their interests *as workers*. Although industrial action in general and strikes in particular are often the main reasons many people hear about trade unions (particularly through the media), most of the work of trade unions is much more mundane. In addition to representing workers who have individual problems at work (such as difficulties over sickness, accidents or pensions) the main function for many trade unions is bargaining with employers over wages and terms and conditions of employment.

Modern trade unions in developed countries are highly institutionalized organizations, with their own staff, elected officers, procedures and internal politics. In some cases they are becoming much more like other member organizations such as clubs and societies, in which services are provided in exchange for a subscription. By no means all trade unions are like this, however, and the origins of trade unions lie in a social movement made up of workers. In many parts of the world, particularly in poorer countries, trade unionism is often much more informal and the social movement nature of the labour movement is much clearer.

The social identities on which the labour movement is based are those associated with employment and social class. As with other identities, 'class' is understood here as (among other things) both relational and (in part) the product of discursive constructions. Because people 'belong' to many social groups developing a sense of *class* identity is not automatic. For many people who have a job, employment is an important part of their self-identity. It is common to hear people describe themselves in terms of their jobs ('I'm a teacher' or 'I'm a secretary' and so on). This sense of occupational identity may form the basis of trade union membership. It is probably less frequent that people describe themselves as members of a class, although it may well be that they do have a sense of their class identity.

Marxian analysts describe class as relational in the sense that it stems from the social relations of production under capitalism. In so far as people have no means of making a living except by selling their labour power for a wage, they exist in a class relation to those who control the means of production. (There is some controversy over the ways in which this abstract relation gets translated into actual social groups, especially in advanced economies.) This is not quite the sense in which Iris Marion Young refers to identity as relational, however. In Young's terms, class is relational, because a sense of class identity depends upon knowing that there are others whose class identity is different. Membership of trade unions and participation in the labour movement depends on at least a minimum sense of occupational identity. It may arise for some people through a sense of class identity too, or a sense of class identity may develop through participation in the labour movement. In all these cases, what class is (and indeed what the labour movement is) is partly the product of discursive construction. Some unions deploy a rhetoric and discourses which stress working-class solidarity. Others emphasize occupational skills and traditions. As with all discourses, labour movements have their heroes, heroines and villains, their narratives of victories and defeats and their symbols of tradition and unity. Furthermore, their discourses do not develop in a vacuum. They intersect with other discourses and identities. For example, some rhetorics of class solidarity and working traditions are markedly gendered masculine (such as those associated with coal mining).

THE RISE OF ORGANIZED LABOUR

Like many other social movements, labour movements are modern phenomena, this time associated particularly with the rise of modern industry. With the development of factories, industrial production and capitalist economies, new kinds of social relations were formed. Traditional ways of life in settled rural communities were disrupted and new working practices, wage relationships and forms of management and business control were developed. This process began first in Britain in the eighteenth century, developing and spreading internationally rapidly through the nineteenth and early twentieth century and forming a global network of capitalist activity in the present day.

Trade unionism and labour movements developed with these changes, and also in relation to the changing character of the state. States sought to regulate the development of trade unionism, channelling it into consideration of economic questions, rather than allowing it to challenge the political order (although it did this too from time to time). They also sought accommodation with labour movements in many places leading to the development of welfare states and, in some cases, incorporation into state decision-making (corporatism). Esping-Andersen's concept of welfare state regimes, which was discussed in Chapter 3, describes in part the process of differential inclusion of the labour movement in different welfare states.

SPACE, PLACE AND LABOUR MOVEMENTS

Understood as a social movement, the labour movement has a complex geography which has only recently become a focus of research for geographers. First, there is not one single labour movement, but many. In the modern world it is common to speak of labour movements in national terms: the British or American labour movement, for example. This reflects the political significance of the modern state and the desire of trade unions to influence government policy. They thus tend to operate at a national level.

To begin with, however, trade unionism tended to be a much more local affair. In Britain craft trade unions, with their roots in the medieval artisanal guilds, were often based in particular towns or cities where a specific craft was practised. At this time the prospect of alliances between workers in different places was one of the things about which both the government and employers were most worried, but tentative steps to broaden the geographical base of support for trade unions were made. The emergence of industrial unions with membership drawn from all trades in a particular industrial sector (coal mining, ship-building, engineering) produced a further degree of geographical integration. A good proportion of the many mergers which occurred among trade unions in the same industry were between unions representing different geographical areas. The twentieth century has seen the rise in many countries of large general unions representing workers not only in different trades and professions, but also in many production sectors. In addition there has been a large increase in the size of the public sector in many countries and in the proportion of its workforce which belong to trade unions. Both of these trends have generated further pressure towards the establishment of national labour movements.

There, have, however, also been developments at the international scale. In the late nineteenth and early twentieth centuries, Marxist politicians in particular argued that the working class should be united across national boundaries. A number of attempts were made to establish socialist 'internationals' which could draw together the strength of labour movements in all capitalist countries. More recently, attempts to develop a single market for goods, capital and labour in Europe have led to calls to establish effective European-level labour organizations to parallel the Europeanization of capital and employment. On the other hand there have also been trends in the opposite direction, with some governments and employers seeking to introduce greater 'flexibility' into the labour market by encouraging, or forcing, trade unions and employees to bargain at a firm, plant, work team or individual level, rather than at a national or industry-wide level.

What is also clear is that there is a complex geography to trade unionism even within one country, industry or trade union. In Britain, for example, trade union membership rates vary widely around the country even within the same industry. A recent study by Ron Martin, Peter Sunley and Jane Wills suggested that, within a pattern of overall decline in trade union membership, some regions of Britain had proved markedly resilient, while others had lost members much more quickly.[17] Such a pattern cannot be accounted for wholly by the socio-economic conditions of different regions, but depends in part on variations in the political strategies, resources and discourses of union organizations in different places. In research into trade unionism in the British public services which I undertook in the late 1980s, I found that trade union responses to the threat of privatization were very different in different parts of the country.[18] These differences were in part a consequence of spatial variation in the privatization process, but were also heavily influenced by the local availability of resources. These included commitment and time on the part of trade union members, representatives and officials, financial resources, organizational infrastructure and local traditions of trade union activity and labour movement culture. In some places it was possible to use a discourse which referred to past activities and represented them as 'the way we do things here'. In other places, with little or no history of campaigning and trade union organizing, there were fewer discursive resources on which local trade unionists could draw.

The geographies of feminism

In contrast to nationalist and labour movements, a variety of social movements which developed particularly strongly during the 1960s and 1970s are often labelled 'new' social movements. As I have noted, this label is something of a misnomer, and it probably applies least of all to the women's movement, which can trace its roots back at least to the late eighteenth century and Mary Wollstonecraft's classic text *The Vindication of the Rights of Woman*, published in 1792. Many contemporary feminist writers have been at pains to show the ways in which women have struggled for civil, social and political rights and equality with men throughout human history and not just since 1960! None the less, although not really new, there is no doubt that the years since the 1960s have seen a large amount of activity, both practical and academic, by women in support of women's rights and in opposition to continuing gender discrimination and inequality. It may not be new, but the women's movement is certainly a social movement in the sense I have outlined here.

FEMINISM IN GEOGRAPHY

Women geographers have participated in growing numbers in the development of feminist ideas and practice. Initially, a major concern was to 'make women visible' to geography which traditionally had been concerned primarily with the spaces and places of men, or which had failed to take account of the extent to which the geographical experiences of the two halves of humanity are markedly and systematically different, most often to the advantage of men. Second, feminist geographers have been exploring the ways in which spatial relations, the character of places and geographical landscapes are both expressive and constitutive of unequal gender relations. What this means is that inequalities between men and women are expressed in the geography of the world (in the social patterning of city life, for example, or in the dominant symbolism of landscapes) but are also influenced by geography (so that the design and planning of cities has an impact on the development of gender relations, for example).[19] More recently still, a number of feminist geographers have been developing ideas about the way in which geographical knowledge is also gendered.[20] Mona Domosh has argued that in the nineteenth century the experiences of women travellers were systematically ignored by the masculine geographical establishment, while those of their male counterparts were elevated to the status of 'scientific' geographical knowledge.[21] Gillian Rose has suggested that the whole tenor of much geographical thought is based on characteristically masculine assumptions about what 'knowing the world' as an academic geographer involves.[22]

While geographers have been engaged for a long time in the study of nationalism, and have begun to develop accounts of the geography of trade unions and the labour movement, there has been, to date, relatively little geographical research on feminism as a social movement. There are some good reasons for this. Many feminist geographers have understandably placed more stress on participating in the women's movement than on writing about it. Moreover, for many feminists, research into the lives and experience of women and their geographies is already (and perhaps necessarily) concerned with women's strategies of resistance to gendered inequalities and masculine oppression, whether that is formalized as a political movement or not. One argument of some feminists suggests that it is in any case inappropriate to judge women's political activity by a masculine norm of what political participation and political organizations involve. While these are important arguments, a study of the geography of social movements should certainly include the women's movement, which has been one of the most influential social movements of the twentieth century.

GEOGRAPHY, DIFFERENCE AND FEMINIST POLITICS

Geography has been involved in the development of feminist politics and of the women's movement in a variety of ways. One important aspect to this is the variation in the development of the women's movement (or rather, women's movements) between different countries. The rapid growth in women's movements during the 1970s began and was initially strongest in developed Western countries. It was organized around campaigns for women's rights in a variety of fields, including the right to equal pay and equal treatment in employment, access to public services, the dramatically unequal distribution of domestic labour and women's disproportionate role in childcare and homemaking (social reproduction), and women's control over their own bodies with regard both to fertility and to male violence.

In the formerly state socialist countries, women were accorded many of the formal rights which their Western counterparts were struggling for, but it has become clear that in many cases such rights only rarely brought gender equality in practice. In the former Soviet bloc, independent women's movements faced similar problems to all attempts to pursue political strategies separately from the state; namely that the effective absence of a developed civil society significantly restricted their sphere of operation. In the poorer countries of the world, women's movements developed around some of the same themes as in the West, but were also particularly concerned with the role of women in 'development', and with the particular impacts on women of severe poverty.

Another (related) set of geographical variations reflect the different experiences of women in different social and cultural systems around the world. Variations in the role of the family, in childcare practices, in cultural attitudes to women and work and in the view of women in different religious traditions have led to important differences in the ways in which women's movements have developed in different places.

In what is sometimes called 'third wave' feminism, these differences in women's experiences began to gain more widespread recognition among feminists. In particular, black feminists stressed the extent to which they were subject to a double discrimination in societies which were racist as well as patriarchal.[23] They also argued that mainstream feminist thinking had not given sufficient weight to the ways in which women in different ethnic, religious and cultural communities experienced 'being women' differently and in ways which did not correspond with a supposed white norm. This heightened awareness of differences between women (which include differences of sexuality, able-bodiedness and class, as well as ethnicity) has raised important questions. One focuses on the extent to which the women's movements should stress the shared oppression of all women as women (and thus perhaps develop a strong sense of gender-based solidarity) or emphasize sensitivity to the particular experiences of groups of women with a variety of identities.

There are also geographical differences in the strength and activities of the women's movement in different parts of the same political system. Susan Halford has investigated the role of initiatives to promote the role and position of women in British local government during the 1980s.[24] She found that there were very marked variations in the extent and character of such initiatives. All the initiatives had been undertaken by local councils controlled by the Labour Party. However, only a minority of Labour councils had developed initiatives. The distribution of women's initiatives was related not only to formal political control, but also to the character of the local Labour Party, the number of women in electorally secure council seats and the social and economic geography of the local area, with women's initiatives more likely in large urban areas, possibly on account of the relatively high proportion of young single women and women in professional occupations.

During the 1980s in the West, the women's movement became closely associated with the peace movement, and in particular with the campaign to end the nuclear arms race and the threat of nuclear war. In the practical pursuit of political strategies, social movements often attempt to disrupt the established rhythm and order of social life. According to Tim Cresswell, in the case of the feminist/peace movement in Britain, this process also involved disrupting the established spatial or geographical order as well.[25] Cresswell argues that the activities of the women peace activists at the missile base at Greenham Common involved (deliberately) transgressing social norms (for example to do with 'feminine' behaviour) and that this transgression was simultaneously a geographical one, which involved undertaking particular activities in the 'wrong' places. As a strategy of subversion this had mixed results, since it also enabled the media, police and local people to represent the Greenham women as 'marginal'.

The futures of social movements

In this chapter I have sought to show how social movements may be interpreted as both the product and the medium of political strategies. As with the state, there can be no all-encompassing theory of social movements which can fully account for all their nuances. Within limits, however, an approach that stresses both the role of resources and preconditions on the one hand, and the importance of the meanings, significance and symbolism which social movements carry on the other, can provide helpful insights. Equally, social movements remain endeavours which are highly porous and prone to fragmentation. As such they are also highly geographically differentiated.

Virtually by definition, the future impacts of social movements are wholly unpredictable. However, in the past they have been absolutely central to many of the most significant changes in the political geography of the world during the twentieth century. Some have argued that social movements have the potential to become the basis of a newly radicalized and participative democracy.[26] Whether this can come to pass remains to be seen, but whatever the future holds, social movements in many ways epitomize the conception of politics I have tried to work with in this book: a politics which is centrally concerned with people, strategies, resources, and change.

Notes to Chapter 6

1 Herbert Blumer, quoted in Anthony Giddens, *The Constitution of Society: Outline of the Theory of Structuration* (Cambridge, Polity Press, 1984), p. 204.
2 Manuel Castells, *The Urban Question* (London, Edward Arnold, 1977); *The City and the Grassroots* (London, Edward Arnold, 1983); S. Lowe, *Urban Social Movements: The City After Castells* (Basingstoke, Macmillan, 1986).
3 Joseph Camilleri and Jim Falk, *The End of Sovereignty?* (Aldershot, Edward Elgar, 1992). See also Graham Smith, 'Political Theory and Human Geography', in Derek Gregory, Ron Martin and Graham Smith, *Human Geography: Society, Space and Social Science* (Basingstoke, Macmillan, 1994), pp. 54–77.
4 Iris Marion Young, *Justice and the Politics of Difference* (Princeton, NJ, Princeton University Press, 1990), p. 158.
5 *Op. cit.*, p. 158.
6 *Op. cit.*, p. 159.
7 *Op. cit.*, p. 160.
8 Gillian Rose, *Feminism and Geography: The Limits of Geographical Knowledge* (Cambridge, Polity Press, 1993), pp. 11–12.
9 Young, *Justice*, p. 172.
10 Alan Scott, *Ideology and the New Social Movements*, (London, Unwin Hyman, 1990); D. Rucht, ed., *Research on Social Movements: The State of the Art in Western Europe and the USA* (Boulder, CO, Westview, 1991).
11 Nigel Thrift, 'On the Determination of Social Action in Space and Time', *Environment and Planning D: Society and Space* 1 (1983), pp. 23–57.
12 I concentrate here on ethnic nationalism. State or official nationalism may be seen as part of the political strategies of states (see Chapters 2 and 3). Another type of nationalism, anti-colonial nationalism, has been considered as part of the anti-colonial strategies discussed in Chapter 4.
13 Tom Nairn, 'The Modern Janus', *New Left Review* 94 (1975), pp. 3–29.
14 Anthony D. Smith, *National Identity* (Harmondsworth, Penguin, 1991), p. 21.
15 Geographers' writings on national identity and nationalism include: James Blaut, *The National Question* (London, Zed Books, 1987); David Hooson, ed., *Geography and National Identity* (Oxford, Blackwell, 1994); Ron Johnston, David Knight and Eleonore Kofman, eds, *Nationalism, Self-determination and Political Geography* (London, Croom Helm, 1988); Colin Williams and Eleonore Kofman, eds, *Community Conflict, Partition and Nationalism* (London, Routledge, 1989).
16 Eric Hobsbawm and Terence Ranger, eds, *The Invention of Tradition* (Cambridge, Cambridge University Press, 1983).

17 Ron Martin, Peter Sunley and Jane Wills, 'The Geography of Trade Union Decline: Spatial Dispersal or Regional Resilience?', *Transactions of the Institute of British Geographers* 18 (1993), pp. 36–62. See also responses by Doreen Massey and by me, and a rejoinder by Martin *et al.* in 'Exchange: Geographies of Trade Unions', *Transactions of the Institute of British Geographers* 19 (1994), pp. 94–118.

18 Joe Painter, 'The Geography of Trade Union Responses to Local Government Privatization', *Transactions of the Institute of British Geographers* 16 (1991), pp. 214–26.

19 Sophie Bowlby, Jane Lewis, Linda McDowell and Jo Foord, 'The Geography of Gender', in Richard Peet and Nigel Thrift, eds, *New Models in Geography Volume Two* (London, Unwin Hyman, 1989), pp. 157–75; Eleonore Kofman and Linda Peake, 'Into the 1990s: A Gendered Agenda for Political Geography', *Political Geography Quarterly* 9 (1990), pp. 313–36; Jo Little, *Gender, Planning and the Policy Process* (Oxford, Pergamon, 1994); Doreen Massey, *Space, Place and Gender*, (Cambridge, Polity Press, 1994); Geraldine Pratt and Susan Hanson, 'Gender, Class and Space', *Environment and Planning D: Society and Space* 6 (1988), pp. 15–35. See also regular 'Progress Reports' on feminist geography and the geography of gender in *Progress in Human Geography* and issues of the journal *Gender, Place and Culture*.

20 Linda McDowell, 'Doing Gender: Feminism, Feminists and Research Methods in Human Geography', *Transactions of the Institute of British Geographers* 17 (1992), pp. 399–416.

21 Mona Domosh, 'Towards a Feminist Historiography of Geography', *Transactions of the Institute of British Geographers* 16 (1991), pp. 95–104.

22 Rose, *Feminism*.

23 bell hooks, *Ain't I a Woman: Black Women and Feminism* (Boston, MA, South End Press, 1981); *Feminist Theory: From Margin to Centre* (Boston, MA, South End Press, 1984).

24 Susan Halford, 'Women's Initiatives in Local Government . . . Where Do They Come From and Where Are They Going?', *Policy and Politics* 16 (1988), pp. 251–59.

25 Tim Cresswell, 'Putting Women in Their Place: The Carnival at Greenham Common', *Antipode: A Radical Journal of Geography* 26 (1994), pp. 35–58.

26 Ernesto Laclau and Chantal Mouffe, *Hegemony and Socialist Strategy* (London, Verso, 1985).

Guide to further reading

Given the range of topics covered there has been space in this book only to introduce the main ideas and some brief examples. Full details of all the works cited in the text is provided in the notes and bibliography for those who are interested. However, you may also find this guide to further reading useful. As well as identifying the most significant and/or accessible works for each topic, it includes a number of additional useful texts to which I have not referred directly in the individual chapters.

Chapter 1

The character of politics is discussed from different perspectives by the contributors to:
Adrian Leftwich, ed., *What is Politics? The Activity and Its Study* (Oxford, Blackwell, 1984)

While some of the problems and paradoxes of contemporary politics are considered by:
Geoff Mulgan, *Politics in an Antipolitical Age* (Cambridge, Polity Press, 1994).

For a flavour of recent developments in Social and Cultural Geography see:
Kay Anderson and Fay Gale, eds, *Cultural Geography: Ways of Seeing* (Melbourne, Longman Cheshire, 1992)
Derek Gregory, *Geographical Imaginations* (Oxford, Blackwell, 1994)
Derek Gregory, Ron Martin and Graham Smith, *Human Geography: Society, Space and Social Science* (Macmillan, Basingstoke, 1994)
Peter Jackson, *Maps of Meaning* (London, Unwin Hyman, 1989)

Chris Philo, compiler, *New Words, New Worlds: Reconceptualising Social and Cultural Geography* (Lampeter, Social and Cultural Geography Study Group, 1991)

Pamela Shurmer-Smith and Kevin Hannam, *Worlds of Desire: Realms of Power* (London, Edward Arnold, 1994)

The interpretative framework I develop in the chapter draws on a variety of sources:

Michèle Barrett, *The Politics of Truth: from Marx to Foucault* (Cambridge, Polity Press, 1991)

Valerie Bryson, *Feminist Political Thought: an Introduction* (Basingstoke, Macmillan, 1992)

Joseph Camilleri and Jim Falk, *The End of Sovereignty?* (Aldershot, Edward Elgar, 1992)

Michel Foucault, 'The Order of Discourse', in Robert Young, ed., *Untying the Text: A Post-structuralist Reader* (London, Routledge and Kegan Paul, 1981), pp. 48–78.

Anthony Giddens, *The Constitution of Society: Outline of the Theory of Structuration* (Cambridge, Polity Press, 1984)

Graham Smith, 'Political Theory and Human Geography' in Derek Gregory, Ron Martin and Graham Smith, *Human Geography: Society, Space and Social Science* (Macmillan, Basingstoke, 1994) pp. 54–77

Nigel Thrift, 'On the Determination of Social Action in Space and Time' *Environment and Planning D: Society and Space* 1 (1983), pp. 23–57

Additional feminist perspectives on politics are provided by:

Judith Butler and Joan W. Scott, eds, *Feminists Theorize the Political* (London, Routledge, 1992)

Carole Pateman, *The Disorder of Women: Democracy, Feminism and Political Theory* (Stanford, CA, Stanford University Press, 1989)

Vicky Randall, *Women and Politics: an International Perspective* (Basingstoke, Macmillan, 1987)

Useful texts which cover the subject matter of political geography include:

Ron Johnston and Peter Taylor, eds, *A World in Crisis: Geographical Perspectives* (Oxford, Blackwell, 1989)

Ron Johnston, Peter Taylor and Michael Watts, eds, *Geographies of Global Change: Remapping the World in the Late Twentieth Century* (Blackwell, Oxford, 1995)

John Short, *An Introduction to Political Geography* (London, Routledge, 2nd edn, 1993)

Peter Taylor *Political Geography: World-economy, Nation-state and Locality* (London, Longman, 1993)

Peter Taylor, ed., *Political Geography of the Twentieth Century* (London, Belhaven Press, 1993)

In addition the journal *Political Geography* is a consistently useful source of new material, and the journal *Progress in Human Geography* contains frequent reviews of recent research and publications in political geography, including regular 'Progress Reports'.

Chapter 2

For an excellent introduction to the modern state have a look at:

Gregor McLennan, David Held and Stuart Hall, eds, *The Idea of the Modern State* (Milton Keynes, Open University Press, 1984)

David Held, 'The Development of the Modern State' in Stuart Hall and Bram Gieben, eds, *Formations of Modernity* (Cambridge, Polity Press, 1992), pp. 71–125.

The concept and the historical processes of state formation are discussed in detail by the following:

Philip Corrigan and Derek Sayer, *The Great Arch: English State Formation as Cultural Revolution* (Oxford, Blackwell, 1985)

Michel Foucault, 'Governmentality', *Ideology and Consciousness* 6 (1979), pp. 5–21

Anthony Giddens, *The Nation-state and Violence* (Cambridge, Polity Press, 1985)

Michael Mann, *States, War and Capitalism* (Oxford, Blackwell, 1988)

Andreas Osiander, *The States System of Europe 1640-1990* (Oxford, Oxford University Press, 1994)

Charles Tilly, *Coercion, Capital and European States: AD 990-1990* (Oxford, Blackwell, 1990)

It is important to contrast the specifics of the European state system with those of other parts of the world. See:

Stuart Corbridge, 'Colonialism, Post-colonialism and the Political Geography of the Third World', in Peter Taylor, ed., *Political Geography of the Twentieth Century* (London, Belhaven Press, 1993)

The case of Africa is considered in detail by:

Jean-François Bayart, *The State in Africa: the Politics of the Belly* (London, Longman, 1993)

The issue of surveillance is central to the modern state. See:

Anthony Giddens, *The Nation-state and Violence* (Cambridge, Polity Press, 1985)

Christopher Dandekar, *Surveillance, Power and Modernity* (Cambridge, Polity Press, 1990)

A key text in debates about the future of the contemporary state is:
Joseph Camilleri and Jim Falk, *The End of Sovereignty?* (Aldershot, Edward Elgar, 1992)

Those debates are followed up through detailed international case studies in the contributions to:
John Dunn, ed., *Contemporary Crisis of the Nation State?* Special Issue of *Political Studies* 42 (1994)

Chapter 3

Key ideas about the organization and operation of the modern state can be explored in more detail in the following:
Patrick Dunleavy and Brenda O'Leary, *Theories of the State* (Basingstoke, Macmillan, 1987)
Anthony Giddens, *The Nation-state and Violence* (Cambridge, Polity Press, 1985)
David Held, 'Central Perspectives on the Modern State', in Gregor McLennan, David Held and Stuart Hall, eds, *The Idea of the Modern State* (Milton Keynes, Open University Press, 1984)
David Held, *Political Theory and the Modern State* (Cambridge, Polity Press, 1984)
Mike Savage and Anne Witz, eds, *Gender and Bureaucracy* (Oxford, Blackwell, 1992)

The 'strategic-relational' approach to the state is outlined in:
Bob Jessop, 'The State as Political Strategy' in *State Theory: Putting Capitalist States in their Place* (Cambridge, Polity Press, 1990), pp. 248-72

The idea of liberalism is discussed by:
John Gray, *Liberalism* (Milton Keynes, Open University Press, 1986)

While the concept of citizenship and its geographies can be followed up in:
G. Andrews, ed., *Citizenship* (London, Lawrence and Wishart, 1991)
J. M. Barbalet, *Citizenship* (Milton Keynes, Open University Press, 1988)
T. H. Marshall, 'Citizenship and Social Class', in T. H. Marshall and Tom Bottomore, *Citizenship and Social Class* (London, Pluto Press, 1991), pp. 3–51
Sallie Marston and Lynn Staeheli, eds, 'Theme Issue: Citizenship', *Environment and Planning A* 26,6 (1994)
Joe Painter and Chris Philo, eds, 'Spaces of Citizenship Special Issue', *Political Geography* 14,2 (1995)

For more detailed information on the welfare state see:

Roger Burrows and Brian Loader, eds, *Towards a Post-Fordist Welfare State?* (London, Routledge, 1994)

Jane Lewis, ed., *Women and Social Policies in Europe: Work, Family and the State* (Aldershot, Edward Elgar, 1993)

Joe Painter, 'The Regulatory State: The Corporate Welfare State and Beyond', in Ron Johnston, Peter Taylor and Michael Watts, eds, *Geographies of Global Change: Remapping the World in the Late Twentieth Century* (Oxford, Blackwell, 1995)

Christopher Pierson, *Beyond the Welfare State* (Cambridge, Polity Press, 1991)

Fiona Williams, *Social Policy: a Critical Introduction* (Cambridge, Polity Press, 1989)

The concept of welfare regimes is summarized by Ron Johnston in:

Ron Johnston, 'The Rise and Decline of the Corporate-welfare State: A Comparative Analysis in Global Context', in Peter Taylor, ed., *Political Geography of the Twentieth Century: a Global Analysis* (London, Belhaven, 1993)

The original text is:

Gøsta Esping-Andersen, *The Three Worlds of Welfare Capitalism* (Cambridge, Polity Press, 1990)

The changing character of the local state and politics is discussed by

Allan Cochrane, *Whatever Happened to Local Government?* (Buckingham, Open University Press, 1993)

Dennis R. Judd and Todd Swanstrom, *City Politics: Private Power and Public Policy* (New York, NY, HarperCollins, 1994)

David Judge, Gerry Stoker and Harold Wolman, eds, *Theories of Urban Politics* (London, Sage, 1995)

Gerry Stoker, *The Politics of Local Government* (Basingstoke, Macmillan, 1988)

David Wilson and Chris Game, *Local Government in the United Kingdom* (Basingstoke, Macmillan, 1994)

The crises of the modern state are discussed in:

Jürgen Habermas, *Legitimation Crisis* (Cambridge, Polity Press, 1988)

James O'Connor, *The Meaning of Crisis* (Oxford, Blackwell, 1987)

Claus Offe, *Contradictions of the Welfare State* (London, Hutchinson, 1984)

Christopher Pierson, *Beyond the Welfare State* (Cambridge, Polity Press, 1991)

An interesting essay on the possible future of the British state is provided by:

Will Hutton, *The State We're In* (London, Cape, 1995)

Chapter 4

For an overview of the topic see:
Diane Elson, 'Imperialism', in Gregor McLennan, David Held and Stuart Hall, eds, *The Idea of the Modern State* (Milton Keynes, Open University Press, 1984)

The histories of European expansion are detailed in the following:
Kenneth R. Andrews, *Trade, Plunder and Settlement* (Cambridge, Cambridge University Press, 1984)
Karl Butzer, ed., *The Americas Before and After 1492: Current Geographical Research*, Special issue of the *Annals of the Association of American Geographers* 82,3 (1992)
D. K. Fieldhouse, *The Colonial Empires: A Comparative Survey from the Eighteenth Century* (Basingstoke, Macmillan, 1981)
V. G. Kiernan, *European Empires from Conquest to Collapse, 1815–1960* (Leicester, Leicester University Press, 1982)
Bernard Porter, *The Lion's Share: A Short History of British Imperialism 1850–1983* (London, Longman, 1984)
Eric R. Wolf, *Europe and the People without History* (Berkeley, CA, University of California Press, 1982)

For discussions of the relationship between geography and imperialism see:
Felix Driver, 'Geography's empire: histories of geographical knowledge', *Environment and Planning D: Society and Space* 10 (1992), pp. 23–40
Anne Godlewska and Neil Smith, eds, *Geography and Empire* (Oxford, Blackwell, 1994)
David Livingstone, *The Geographical Tradition: Episodes in the History of a Contested Enterprise* (Oxford, Blackwell, 1992)

The ideas of Wallerstein's world-systems theory are outlined in:
Peter Taylor, *Political Geography: World-economy, Nation-state and Locality* (London, Longman, 1993). See especially Chapter 1.

An accessible introduction by Wallerstein himself is:
Immanuel Wallerstein, *Historical Capitalism* (London, Verso, 1983)

The diverse strategies of colonial domination (with particular reference to cultural and discursive practices) are discussed in the following texts:
Anton Gill, *Ruling Passions: Sex, Race and Empire* (London, BBC Books, 1995)
Derek Gregory, *Geographical Imaginations* (Oxford, Blackwell, 1994)
Stuart Hall, 'The West and the Rest: Discourse and Power' in Stuart Hall and Bram Gieben, eds, *Formations of Modernity* (Cambridge, Polity Press, 1992), pp. 275–331

J. A. Mangan, ed., *Making Imperial Mentalities: Socialisation and British Imperialism* (Manchester, Manchester University Press, 1990)

Timothy Mitchell, *Colonizing Egypt* (Cambridge, Cambridge University Press, 1988)

Edward Said, *Orientalism: Western Conceptions of the Orient* (Harmondsworth, Penguin, 1978)

Edward Said, *Culture and Imperialism* (London, Chatto and Windus, 1993)

Robert Young, *White Mythologies: Writing History and the West* (London, Routledge, 1990)

On post-colonialism see:

Stuart Corbridge, 'Colonialism, Post-colonialism and the Political Geography of the Third World', in Peter Taylor, ed., *Political Geography of the Twentieth Century* (London, Belhaven Press, 1993)

Jonathan Crush, 'Post-colonialism, De-colonization, and Geography', in Anne Godlewska and Neil Smith, eds, *Geography and Empire* (Oxford, Blackwell, 1994)

Derek Gregory, *Geographical Imaginations* (Oxford, Blackwell, 1994). See especially pp. 165–203.

Chapter 5

On the histories of the 'Cold War' and its aftermath see:

Simon Dalby, *Creating the Second Cold War: The Discourse of Politics* (London, Frances Pinter, 1990)

Charles W. Kegley Jr. and Gregory Raymond, *A Multipolar Peace? Great Power Politics in the Twenty-first Century* (New York, St Martin's Press, 1994)

Daniel S. Papp, *Contemporary International Relations: Frameworks for Understanding* (New York, NY, Macmillan Publishing Company, 1991)

Richard Sakwa, *Gorbachev and His Reforms, 1985-1990* (Hemel Hempstead, Philip Allan, 1990)

Peter Taylor, *Britain and the Cold War: 1945 as Geopolitical Transition* (London, Pinter, 1990)

Contrasting approaches to international relations and geopolitics are surveyed in:

Richard Little and Michael Smith, *Perspectives on World Politics* (London, Routledge, 1991)

On 'geopolitical economy' see:

John Agnew and Stuart Corbridge, *Mastering Space: Hegemony, Territory and International Political Economy* (London, Routledge, 1995)

On critical geopolitics see:

Simon Dalby, 'Critical Geopolitics: Discourse, Difference and Dissent' *Environment and Planning D: Society and Space* 9 (1991) pp. 261–83

Simon Dalby, 'Gender and Critical Geopolitics: Reading Security Discourse in the New World Disorder', *Environment and Planning D: Society and Space* 12 (1994), pp. 595–612

Klaus-John Dodds and James Sidaway, 'Locating Critical Geopolitics', *Environment and Planning D: Society and Space* 12 (1994), pp. 515–24

Jim George, *Discourses of Global Politics: A Critical (Re)Introduction to International Relations* (Boulder, CO, Rienner, 1994)

For examples of the role of discourse analysis in the interpretation of international relations have a look at:

Miriam Cooke and Angela Woollacott, *Gendering War Talk* (Princeton, NJ, Princeton University Press, 1993)

Cynthia Enloe, *The Morning After: Sexual Politics at the End of the Cold War* (Berkeley, University of California Press, 1993)

Gearóid Ó Tuathail and John Agnew, 'Geopolitics and Discourse: Practical Geopolitical Reasoning in American Foreign Policy', *Political Geography* 11 (1992), pp. 190–204

Stephen J. Whitfield, *The Culture of the Cold War* (Baltimore, MD, Johns Hopkins University Press, 1991)

Chapter 6

The ideas in this chapter are discussed in much more detail in the following:

Ernesto Laclau and Chantal Mouffe, *Hegemony and Socialist Strategy* (London, Verso, 1985)

Stanford M. Lyman, ed., *Social Movements: Critiques, Concepts, Case Studies* (Basingtoke, Macmillan, 1995)

D. Rucht, ed., *Research on Social Movements: the State of the Art in Western Europe and the USA* (Boulder, CO, Westview, 1991)

Alan Scott, *Ideology and the New Social Movements* (London, Unwin Hyman, 1990)

Iris Marion Young, *Justice and the Politics of Difference* (Princeton University Press, Princeton, NJ, 1990)

Concrete examples of a variety of contemporary social movements around the world are described in:

Paul Ekins, *A New World Order: Grassroots Movements for Global Change* (London, Routledge, 1992)

On the geographies of ethnic nationalism see:
Peter Alter, *Nationalism* (London, Edward Arnold, 1994)
James Blaut, *The National Question* (London, Zed Books, 1987)
David Hooson, ed., *Geography and National Identity* (Oxford, Blackwell, 1994)
Anthony D. Smith, *National Identity* (Harmondsworth, Penguin, 1991)
Graham Smith, ed., *The Nationalities Question in the Soviet Union* (London, Longman, 1990)

On the geography of British labour movements see:
Mick Griffiths and Ron Johnston, 'What's in a Place? An Approach to the Concept of Place, as Illustrated by the British National Union of Mineworker' strike, 1984–5', *Antipode* 23 (1991), pp. 185–213.
Ron Martin, Peter Sunley and Jane Wills, 'The Geography of Trade Union Decline: Spatial Dispersal or Regional Resilience?' *Transactions of the Institute of British Geographers* 18 (1993), pp. 36–62
Joe Painter, 'The Geography of Trade Union Responses to Local Government Privatization', *Transactions of the Institute of British Geographers* 16 (1991), pp. 214–26
Humphrey Southall, 'Towards a Geography of Unionisation: The Spatial Organisation and Distribution of the Early British Trade Unions' *Transactions of the Insitute of British Geographers* 13 (1988), pp. 467–83

And for the United States:
Gordon Clark, *Unions and Communities Under Siege* (Cambridge, Cambridge University Press, 1990)

A key text on the relationship between geography and feminism is:
Gillian Rose, *Feminism and Geography: The Limits of Geographical Knowledge* (Cambridge, Polity Press, 1993)

A range of concrete examples of the growth and activities of women's movements in different parts of the world are discussed in the following:
Susan Bassnett, *Feminist Experiences: the Women's Movement in Four Cultures* (London, Allen and Unwin, 1986)
Sarah A. Radcliffe and Sallie Westwood, *Viva! Women and Popular Protest in Latin American* (London, Routledge, 1993)
Sheila Rowbotham, *Women in Movement: Feminism and Social Action* (London, Routledge, 1992)

Bibliography

Agnew, John and Stuart Corbridge, 'The New Geopolitics: The Dynamics of Geopolitical Disorder', in Ron Johnston and Peter Taylor, eds, *A World in Crisis: Geographical Perspectives* (Oxford, Blackwell, 1989), pp. 266–88

Alter, Peter, *Nationalism* (London, Edward Arnold, 1994)

Anderson, Kay and Fay Gale, eds, *Cultural Geography: Ways of Seeing* (Melbourne, Longman Cheshire, 1992)

Andrews, Kenneth R., *Trade, Plunder and Settlement* (Cambridge, Cambridge University Press, 1984)

Archer, J. C., 'Public Choice Paradigms in Political Geography', in A. D. Burnett and P. J. Taylor, eds, *Political Studies from Spatial Perspectives* (New York, John Wiley, 1981)

Bacon, Roger and Walter Eltis, *Britain's Economic Problem: Too Few Producers* (London, Macmillan, 1978)

Bakshi, Parminder, Mark Goodwin, Joe Painter and Alan Southern, 'Gender, Race and Class in the Local Welfare State', *Environment and Planning A* (forthcoming)

Barrett, Michèle, *The Politics of Truth: From Marx to Foucault* (Cambridge, Polity Press, 1991)

Bayart, Jean-François, *The State in Africa: The Politics of the Belly* (London, Longman, 1993)

Bell, David, 'Pleasure and Danger: The Paradoxical Spaces of Sexual Citizenship', *Political Geography* 14 (1995), pp. 139–53

Bennett, Robert, *The Geography of Public Finance* (London, Methuen, 1980)

Berman, Marshall, *All That is Solid Melts into Air* (London, Verso, 1982)

Blaut, James, *The National Question* (London, Zed Books, 1987)

Boorstin, Daniel J., *The Discoverers* (Harmondsworth, Penguin, 1986)

Bowlby, Sophie, Jane Lewis, Linda McDowell and Jo Foord, 'The Geography of Gender', in Richard Peet and Nigel Thrift, eds, *New*

Models in Geography Volume Two (London, Unwin Hyman, 1989), pp. 157–75

Bryson, Valerie, *Feminist Political Thought: An Introduction* (Basingstoke, Macmillan, 1992)

Burrows, Roger and Brian Loader, eds, *Towards a Post-Fordist Welfare State?* (London, Routledge, 1994)

Camilleri, Joseph and Jim Falk, *The End of Sovereignty?* (Aldershot, Edward Elgar, 1992)

Carter, Paul, *The Road to Botany Bay* (London, Faber and Faber, 1987)

Castells, Manuel, *The Urban Question* (London, Edward Arnold, 1977)

Castells, Manuel, *The City and the Grassroots* (London, Edward Arnold, 1983)

Chatterjee, Partha, *Nationalist Thought and the Colonial World: A Derivative Discourse?* (London, Zed Books, 1986)

Clark, Gordon and Michael Dear, *State Apparatus: Structures and Language of Legitimacy* (Boston, MA, Allen and Unwin, 1984)

Clarke, J., A. Cochrane and C. Smart, *Ideologies of Welfare* (London, Hutchinson, 1987)

Clayton, Anthony, *The Wars of French Decolonization* (London, Longman, 1994)

Cochrane, Allan, *Whatever Happened to Local Government?* (Buckingham, Open University Press, 1993)

Cockburn, Cynthia, *The Local State* (London, Pluto Press, 1977)

Cooke, Miriam, and Angela Woollacott, *Gendering War Talk* (Princeton, NJ, Princeton University Press, 1993)

Corbridge, Stuart, 'Colonialism, Post-colonialism and the Political Geography of the Third World', in Peter Taylor, ed., *Political Geography of the Twentieth Century* (London, Belhaven Press, 1993), pp. 171–205

Corbridge, Stuart, 'Marxisms, Modernities, and Moralities: Development Praxis and the Claims of Distant Strangers', *Environment and Planning D: Society and Space* 11 (1993), pp. 449–72

Corrigan, Philip and Derek Sayer, *The Great Arch: English State Formation as Cultural Revolution* (Oxford, Blackwell, 1985)

Cosgrove, Denis and Stephen Daniels, eds, *The Iconography of Landscape* (Cambridge, Cambridge University Press, 1988)

Cresswell, Tim, 'Putting Women in Their Place: The Carnival at Greenham Common', *Antipode: A Radical Journal of Geography* 26 (1994), pp. 35–58

Crush, Jonathan, 'Post-colonialism, De-colonization, and Geography', in Anne Godlewska and Neil Smith, eds, *Geography and Empire* (Oxford, Blackwell, 1994)

Curtis, Sarah, *The Geography of Public Welfare Provision* (London, Routledge, 1989)

Dalby, Simon, *Creating the Second Cold War: The Discourse of Politics* (London, Frances Pinter, 1990)

Dalby, Simon, 'Critical Geopolitics: Discourse, Difference and Dissent', *Environment and Planning D: Society and Space* 9 (1991) pp. 261–83

Dalby, Simon, 'Gender and Critical Geopolitics: Reading Security Discourse in the New World Disorder', *Environment and Planning D: Society and Space* 12 (1994), pp. 595–612

Dandekar, Christopher, *Surveillance, Power and Modernity* (Cambridge, Polity Press, 1990)

De Schweinitz Jr, Karl, *The Rise and Fall of British India: Imperialism as Inequality* (London, Methuen, 1983)

Dear, Michael, 'The Postmodern Challenge: Reconstructing Human Geography', *Transactions of the Institute of British Geographers* 13 (1988), pp. 262–74

Dodds, Klaus-John, 'Geopolitics and Foreign Policy: Recent Developments in Anglo-American Political Geography and International Relations', *Progress in Human Geography* 18 (1994), pp. 186–208

Dodds, Klaus-John and James Sidaway, 'Locating Critical Geopolitics', *Environment and Planning D: Society and Space* 12 (1994), pp. 515–24

Domosh, Mona, 'Towards a Feminist Historiography of Geography', *Transactions of the Institute of British Geographers* 16 (1991), pp. 95–104

Driver, Felix, 'Geography's Empire: Histories of Geographical Knowledge', *Environment and Planning D: Society and Space* 10 (1992), pp. 23–40

Driver, Felix, *Power and Pauperism: The Workhouse System 1834–1884* (Cambridge, Cambridge University Press, 1993)

Drover, Glenn and Patrick Kerans, eds, *New Approaches to Welfare Theory* (Aldershot, Edward Elgar, 1993)

Duncan, Simon and Mark Goodwin, *The Local State and Uneven Development* (Cambridge, Polity Press, 1988)

Ekins, Paul, *A New World Order: Grassroots Movements for Global Change* (London, Routledge, 1992)

Elson, Diane, 'Imperialism', in Gregor McLennan, David Held and Stuart Hall, eds, *The Idea of the Modern State* (Milton Keynes, Open University Press, 1984), pp. 154–82

Enloe, Cynthia, *The Morning After: Sexual Politics at the End of the Cold War* (Berkeley, CA, University of California Press, 1993)

Esping-Andersen, Gøsta, *The Three Worlds of Welfare Capitalism* (Cambridge, Polity Press, 1990)

Fieldhouse, D. K., *The Colonial Empires: A Comparative Survey from the Eighteenth Century* (Basingstoke, Macmillan, 1981)

Foucault, Michel, 'Governmentality', *Ideology and Consciousness* 6 (1979), pp. 5–21

Foucault, Michel, 'The Order of Discourse', in Robert Young, ed., *Untying the Text: Post-structuralist Reader* (London, Routledge and Kegan Paul, 1981), pp. 48–78

Frank, A. G., *Capitalism and Underdevelopment in Latin America* (New York, Monthly Review Press, 1969)

Fukuyama, Francis, 'The End of History?' *The National Interest* (Summer, 1989)

Fukuyama, Francis, *The End of History and the Last Man* (New York, Free Press, 1992)

George, Jim, *Discourses of Global Politics: A Critical (Re)Introduction to International Relations* (Boulder, CO, Rienner, 1994)

Gibbs, Philip, *The Romance of Empire* (London, Hutchinson, undated)

Giddens, Anthony, *The Constitution of Society: Outline of the Theory of Structuration* (Cambridge, Polity Press, 1984)

Giddens, Anthony, *The Nation-state and Violence* (Cambridge, Polity Press, 1985)

Gill, Anton, *Ruling Passions: Sex, Race and Empire* (London, BBC Books, 1995).

Glassner, Martin Ira, *Political Geography* (New York, John Wiley and Sons, 1993)

Godlewska, Anne and Neil Smith, eds, *Geography and Empire* (Oxford, Blackwell, 1994)

Gray, John, *Liberalism* (Milton Keynes, Open University Press, 1986)

Gregory, Derek, *Geographical Imaginations* (Oxford, Blackwell, 1994)

Gregory, Derek, Ron Martin and Graham Smith, *Human Geography: Society, Space and Social Science* (Basingstoke, Macmillan, 1994)

Griffith Taylor, T., 'Racial Geography', in T. Griffith Taylor, ed., *Geography in the Twentieth Century* (New York, Philosophical Library, 3rd edn, 1957)

Habermas, Jürgen, *Legitimation Crisis* (Cambridge, Polity Press, 1988)

Halford, Susan, 'Women's Initiatives in Local Government ... Where Do They Come From and Where Are They Going?', *Policy and Politics* 16 (1988), pp. 251–59.

Hall, Stuart, 'The West and the Rest: Discourse and Power', in Stuart Hall and Bram Gieben, eds, *Formations of Modernity* (Cambridge, Polity Press, 1992), pp. 275–331

Hartshorne, Richard, *The Nature of Geography. A Critical Survey of Current Thought in the Light of the Past* (Lancaster, PA, Association of American Geographers, 1939)

Harvey, David, *The Limits to Capital* (Oxford, Blackwell, 1982)

Hepple, Leslie, 'Destroying Local Leviathans and Designing Landscapes of Liberty? Public Choice Theory and The Poll Tax', *Transactions of the Institute of British Geographers* 14 (1989), pp. 387–99

Hobsbawm, Eric and Terence Ranger, eds, *The Invention of Tradition* (Cambridge, Cambridge University Press, 1983)

hooks, bell, *Ain't I a Woman: Black Women and Feminism* (Boston, MA, South End Press, 1981)

hooks, bell, *Feminist Theory: From Margin to Centre* (Boston, MA, South End Press, 1984)

Hooson, David, ed., *Geography and National Identity* (Oxford, Blackwell, 1994)

Hutton, Will, *The State We're In* (London, Cape, 1995)

Ingham, Geoffrey, *Capitalism Divided: The City and Industry in British Social Development* (Basingstoke, Macmillan, 1984)

Jackson, Peter, *Maps of Meaning* (London, Unwin Hyman, 1989)

Jessop, Bob, 'The State as Political Strategy', in *State Theory: Putting Capitalist States in Their Place* (Cambridge, Polity Press, 1990), pp. 248–72

Jessop, Bob, 'The Transition to Post-Fordism and The Schumpeterian Workfare State' in Roger Burrows and Brian Loader, eds, *Towards a Post-Fordist Welfare State?* (London, Routledge, 1994)

Johnston, Ron, *Geography and the State: An Essay in Political Geography* (Basingstoke, Macmillan, 1982)

Johnston, Ron, 'The Rise and Decline of The Corporate-Welfare State: A Comparative Analysis in Global Context', in Peter Taylor, ed., *Political Geography of the Twentieth Century: A Global Analysis* (London, Belhaven, 1993)

Johnston, Ron, David Knight and Eleonore Kofman, eds, *Nationalism, Self-determination and Political Geography* (London, Croom Helm, 1988)

Johnston, Ron, Fred Shelley and Peter Taylor, eds, *Developments in Electoral Geography* (London, Routledge, 1990)

Johnston, Ron and Peter Taylor, eds, *A World in Crisis: Geographical Perspectives* (Oxford, Blackwell, 1989)

Johnston, Ron, Peter Taylor and Michael Watts, eds, *Geographies of Global Change: Remapping the World in the Late Twentieth Century* (Oxford, Blackwell, 1995)

Judd, Dennis R. and Todd Swanstrom, *City Politics: Private Power and Public Policy* (New York, HarperCollins, 1994)

Kariuki, Josiah Mwangi, 'The "Mau-Mau" Oath', in Elie Kedourie, ed., *Nationalism in Asia and Africa* (London, Weidenfeld and Nicolson, 1970)

Kegley, Charles W., Jr and Gregory Raymond, *A Multipolar Peace? Great Power Politics in the Twenty-first Century* (New York, St Martin's Press, 1994)

Kennan, George F. ('Mr X'), 'The Sources of Soviet Conduct', *Foreign Affairs* 25 (1947), pp. 566–82

Kiernan, V. G., *European Empires from Conquest to Collapse, 1815–1960* (Leicester, Leicester University Press, 1982)

Kofman, Eleonore, 'Citizenship for Some but Not for Others: Spaces of Citizenship in Contemporary Europe', *Political Geography* 14 (1995), pp. 121–37

Kofman, Eleonore and Linda Peake, 'Into the 1990s: A Gendered Agenda for Political Geography', *Political Geography Quarterly* 9 (1990), pp. 313–36

Laclau, Ernesto and Chantal Mouffe, *Hegemony and Socialist Strategy* (London, Verso, 1985).

Leftwich, Adrian, ed., *What is Politics? The Activity and Its Study* (Oxford, Blackwell, 1984)

Lewis, Jane, ed., *Women and Social Policies in Europe: Work, Family and the State* (Aldershot, Edward Elgar, 1993)

Lipset S. M. and Stein Rokkan, 'Cleavage Structures, Party Systems and Voter Alignments', in S. M. Lipset and Stein Rokkan, eds, *Party Systems and Voter Alignments: Cross-national Perspectives* (New York, Free Press, 1967)

Little, Jo, *Gender, Planning and the Policy Process* (Oxford, Pergamon, 1994)

Livingstone, David, *The Geographical Tradition: Episodes in the History of a Contested Enterprise* (Oxford, Blackwell, 1992)

Lowe, S., *Urban Social Movements: The City After Castells* (Basingstoke, Macmillan, 1986)

Mackinder, Halford, 'The Geographical Pivot of History', *The Geographical Journal* 23 (1904), pp. 421–42

Mackinder, Halford, *Democratic Ideals and Reality* (London, Constable, 1919)

Mangan, J. A., ed., *Making Imperial Mentalities: Socialisation and British Imperialism* (Manchester, Manchester University Press, 1990)

Mann, Michael, 'The Autonomous Power of the State: Its Origins, Mechanisms and Results', in *States, War and Capitalism* (Oxford, Blackwell, 1988)

Mann, Michael, *States, War and Capitalism* (Oxford, Blackwell, 1988)

Marshall, T. H., *Citizenship and Social Class* (Cambridge, Cambridge University Press, 1950)

Marshall, T. H., 'Citizenship and Social Class', in T. H. Marshall and Tom Bottomore, *Citizenship and Social Class* (London, Pluto Press, 1991), pp. 3–51

Marston, Sallie and Lynn Staeheli, eds, 'Theme Issue: Citizenship', *Environment and Planning A* 26,6 (1994)

Martin, Ron, Peter Sunley and Jane Wills, 'The Geography of Trade Union Decline: Spatial Dispersal or Regional Resilience?', *Transactions of the Institute of British Geographers* 18 (1993), pp. 36–62

Massey, Doreen, 'New Directions in Space', in Derek Gregory and John Urry, eds, *Social Relations and Spatial Structures* (Basingstoke, Macmillan, 1985), pp. 9–19

Massey, Doreen, *Space, Place and Gender* (Cambridge, Polity Press, 1994)

McDowell, Linda, 'Doing Gender: Feminism, Feminists and Research Methods in Human Geography', *Transactions of the Institute of British Geographers* 17 (1992), pp. 399–416

McDowell, Linda, 'The Transformation of Cultural Geography', in Derek Gregory, Ron Martin and Graham Smith, eds, *Human Geography: Society, Space and Social Science* (Basingstoke, Macmillan, 1994), pp. 146–73

McLennan, Gregor, David Held and Stuart Hall, eds, *The Idea of the Modern State* (Milton Keynes, Open University Press, 1984)

Mitchell, Timothy, *Colonizing Egypt* (Cambridge, Cambridge University Press, 1988)

Mulgan, Geoff, *Politics in an Antipolitical Age* (Cambridge, Polity Press, 1994)

Nairn, Tom, 'The Modern Janus', *New Left Review* 94 (1975), pp. 3–29

Nairn, Tom, *The Enchanted Glass: Britain and its Monarchy* (London, Radius, 1988)

Ó Tuathail, Gearóid, 'Putting Mackinder in His Place', *Political Geography* 11 (1992), pp. 100–18

Ó Tuathail, Gearóid, '(Dis)placing Geopolitics: Writing on the Maps of Global Politics' *Environment and Planning D: Society and Space* 12 (1994) pp. 525–46

Ó Tuathail, Gearóid, 'Problematizing Geopolitics: Survey, Statesmanship and Strategy', *Transactions of the Institute of British Geographers* 19 (1994), pp. 259–72

Ó Tuathail, Gearóid and John Agnew, 'Geopolitics and Discourse: Practical Geopolitical Reasoning in American Foreign Policy', *Political Geography* 11 (1992), pp. 190–204

Ó Tuathail, Gearóid and T. Luke, 'Present at (Dis)integration: Deterritorialization and Reterritorialization in the New Wor(l)d Order', *Annals of the Association of American Geographers* 84 (1994), pp. 381–98

O'Connor, James, *The Fiscal Crisis of the State* (New York, St Martin's Press, 1973)

O'Connor, James, *The Meaning of Crisis* (Oxford, Blackwell, 1987)

Offe, Claus, *Contradictions of the Welfare State* (London, Hutchinson, 1984)

Ogborn, Miles, 'Local Power and State Regulation in Nineteenth Century Britain', *Transactions of the Institute of British Geographers* 17 (1992), pp. 215–26

Painter, Joe, 'The Geography of Trade Union Responses to Local Government Privatization', *Transactions of the Institute of British Geographers* 16 (1991), pp. 214–26

Painter, Joe, 'The Culture of Competition', *Public Policy and Administration*, 7,1 (1992), pp. 58–68

Painter, Joe and Chris Philo, eds, 'Spaces of Citizenship Special Issue', *Political Geography* 14,2 (1995)

Papp, Daniel S., *Contemporary International Relations: Frameworks for Understanding* (New York, Macmillan, 1991)

Pateman, Carole, *The Disorder of Women: Democracy, Feminism and Political Theory* (Stanford, CA, Stanford University Press, 1989)

Philo, Chris, ' "Fit Localities for an Asylum": The Historical Geography of the Nineteenth-Century "Mad-business" in England as Viewed Through the Pages of the *Asylum Journal*', *Journal of Historical Geography* 13 (1987), pp. 398–415

Philo, Chris, compiler, *New Words, New Worlds: Reconceptualising Social and Cultural Geography* (Lampeter, Social and Cultural Geography Study Group, 1991)

Pierson, Christopher, *Beyond the Welfare State* (Cambridge, Polity Press, 1991)

Pincetl, Stephanie, 'Challenges to Citizenship: Latino Immigrants and Political Organizing in the Los Angeles Area', *Environment and Planning A* 26 (1994), pp. 895–914

Pinch, Steven, *Cities and Services* (London, Routledge and Kegan Paul, 1980)

Poulantzas, Nicos, *State, Power, Socialism* (London, New Left Books, 1978)

Pratt, Geraldine and Susan Hanson, 'Gender, Class and Space', *Environment and Planning D: Society and Space* 6 (1988), pp. 15–35

Ratzel, Friedrich, 'The Laws of the Spatial Growth of States', in Roger Kasperson and Julian Minghi, eds, *The Structure of Political Geography* (London, London University Press, 1969), pp. 17–28

Rokkan, Stein, *Citizens, Elections and Parties* (New York, McKay, 1970)

Rose, Gillian, *Feminism and Geography: The Limits of Geographical Knowledge* (Cambridge, Polity Press, 1993)

Rucht, D., ed., *Research on Social Movements: The State of the Art in Western Europe and the USA* (Boulder, CO, Westview, 1991)

Said, Edward, *Orientalism: Western Conceptions of the Orient* (Harmondsworth, Penguin, 1978)

Said, Edward, *Culture and Imperialism* (London, Chatto and Windus, 1993)

Sakwa, Richard, *Gorbachev and His Reforms, 1985–1990* (Hemel Hempstead, Philip Allan, 1990)

Savage, Mike, and Anne Witz, eds, *Gender and Bureaucracy* (Oxford, Blackwell, 1992)

Scott, Alan, *Ideology and the New Social Movements* (London, Unwin Hyman, 1990)

Short, John, *An Introduction to Political Geography* (London, Routledge, 2nd edn 1993)

Shurmer-Smith, Pamela and Kevin Hannam, *Worlds of Desire: Realms of Power* (London, Edward Arnold, 1994)

Smith, Anthony D., *National Identity* (Harmondsworth, Penguin, 1991)

Smith, David M., *Geography and Social Justice* (Oxford, Blackwell, 1994)

Smith, Graham, ed., *The Nationalities Question in the Soviet Union* (London, Longman, 1990)

Smith, Graham, 'Political Theory and Human Geography', in Derek Gregory, Ron Martin and Graham Smith, *Human Geography: Society, Space and Social Science* (Macmillan, Basingstoke, 1994), pp. 54–77

Smith, Michael Peter, *City State and Market: The Political Economy of Urban Society* (Oxford, Blackwell, 1988)

Smith, Neil, *Uneven Development* (Oxford, Blackwell, 1984)

Soja, Edward, 'Communications and Territorial Integration in East Africa' *East Lakes Geographer* 4 (1968), pp. 39–57

Soja, Edward, 'The Socio-spatial Dialectic', *Annals of the Association of American Geographers* 70 (1980), pp. 207–25

Soja, Edward, *Postmodern Geographies: The Reassertion of Space in Critical Social Theory* (London, Verso, 1989)

Spivak, Gayatri Chakravorty, 'Can the Subaltern Speak?', in Cary Nelson and Lawrence Grossberg, eds, *Marxism and the Interpretation of Culture* (Chicago, University of Illinois Press, 1988), pp. 271–313

Stoddart, David, *On Geography and Its History* (Oxford, Blackwell, 1986)

Stoker, Gerry, *The Politics of Local Government* (Basingstoke, Macmillan, 1988)

Suny, Ronald, 'Incomplete Revolution: National Movements and the Collapse of the Soviet Empire', *New Left Review* 189 (1991), pp. 111–25

Taylor, Peter, *Political Geography: World-economy, Nation-state and Locality* (London, Longman, 1989)

Taylor, Peter, 'The Error of Developmentalism in Human Geography', in Derek Gregory and Rex Walford, eds, *Horizons in Human Geography* (Basingstoke, Macmillan, 1989), pp. 303–19

Taylor, Peter, ed., *Political Geography of the Twentieth Century* (London, Belhaven Press, 1993)

Taylor, Peter, 'Political Geography', in Ron Johnston, Derek Gregory and David Smith, eds, *The Dictionary of Human Geography* (Oxford, Blackwell, 1994)

Thiong'o, Ngugi wa, *Decolonizing the Mind* (London, James Currey, 1986)

Thrift, Nigel, 'On the Determination of Social Action in Space and Time', *Environment and Planning D: Society and Space* 1 (1983), pp. 23–57

Tiebout, C. M., 'A Pure Theory of Local Expenditures', *Journal of Political Economy* 64 (1956), pp. 416–24

Tilly, Charles, *Coercion, Capital and European states: AD 990–1990* (Oxford, Blackwell, 1990)

Whitfield, Stephen J., *The Culture of the Cold War* (Baltimore, MD, Johns Hopkins University Press, 1991)

Williams, Colin and Eleonore Kofman, eds, *Community Conflict, Partition and Nationalism* (London, Routledge, 1989)

Williams, Fiona, *Social Policy: A Critical Introduction* (Cambridge, Polity Press, 1989)

Wilson, David and Chris Game, *Local Government in the United Kingdom* (Basingstoke, Macmillan, 1994)

Wolch, Jennifer, 'The Shadow State: Transformations in the Voluntary Sector', in Jennifer Wolch and Michael Dear, eds, *The Power of Geography* (Boston, MA, Unwin Hyman, 1989), pp. 197–221

Wolf, Eric R., *Europe and the People without History* (Berkeley, CA, University of California Press, 1982), pp. 131–57

Young, Iris Marion, *Justice and the Politics of Difference* (Princeton, NJ, Princeton University Press, 1990)

Young, Robert, *White Mythologies: Writing History and the West* (London, Routledge, 1990)

Index